T0292615

Springer Tracts in Modern Physics

Volume 267

Springer Tracts in Modern Physics

Springer Tracts in Modern Physics provides comprehensive and critical reviews of topics of current interest in physics. The following fields are emphasized: Elementary Particle Physics, Condensed Matter Physics, Light Matter Interaction, Atomic and Molecular Physics, Complex Systems, Fundamental Astrophysics.

Suitable reviews of other fields can also be accepted. The editors encourage prospective authors to correspond with them in advance of submitting a manuscript. For reviews of topics belonging to the above mentioned fields, they should address the responsible editor as listed in "Contact the Editors".

Condensed Matter Physics
Yan Chen
Fudan University
Department of Physics
2250 Songhu Road
Shanghai, China 400438
Email: yanchen99@fudan.edu.cn
www.physics.fudan.edu.cn/tps/branch/cqc/en/
people/faculty/

Atsushi Fujimori
Editor for The Pacific Rim
Department of Physics
University of Tokyo
7-3-1 Hongo, Bunkyo-ku
Tokyo 113-0033, Japan
Email: fujimori@phys.s.u-tokyo.ac.jp
http://wyvern.phys.s.u-tokyo.ac.jp/welcome_en.
html

Atomic, Molecular and Optical Physics
William C. Stwalley
University of Connecticut
Department of Physics
2152 Hillside Road, U-3046
Storrs, CT 06269-3046, USA
Email: w.stwalley@uconn.edu
www-phys.uconn.edu/faculty/stwalley.html

Solid State and Optical Physics
Ulrike Woggon
Institut für Optik und Atomare Physik
Technische Universität Berlin
Straße des 17. Juni 135
10623 Berlin, Germany
Email: ulrike.woggon@tu-berlin.de
www.ioap.tu-berlin.de

More information about this series at http://www.springer.com/series/426

Matthias U. Mozer

Electroweak Physics
at the LHC

 Springer

Matthias U. Mozer
Institut für Experimentelle Kernphysik
KIT
Karlsruhe, Baden-Württemberg
Germany

ISSN 0081-3869   ISSN 1615-0430   (electronic)
Springer Tracts in Modern Physics
ISBN 978-3-319-30380-2   ISBN 978-3-319-30381-9   (eBook)
DOI 10.1007/978-3-319-30381-9

Library of Congress Control Number: 2016932002

This Springer imprint is published by Springer Nature
The registered company is Springer International Publishing AG Switzerland

Acknowledgements

First and foremost I would like to thank Thomas Müller, who has made this book possible in the first place. Without his continued support my work would not have been possible.

I'd also like to extend my gratitude to Pietro Govoni, Maurizio Pierini and Gautier Hamel de Monchenault for reviews of the draft and useful discussions. Additionally, I'd like to thank Matthew Herndon for his assistance in the preparation of summary plots.

The book would have been impossible without the hard work of the people at CERN, who built and operated the LHC with immense success. Similarly, I'd like to thank my colleagues in the CMS and ATLAS collaboration for building and operating the detectors as well as for their untiring efforts to use the collected data in meaningful physics measurements, which I had the honour to summarise in this book.

I'd like to thank CERN as well as the CMS and ATLAS collaborations for their permissions to reuse their figures. Similarly I am indebted to the LEP electroweak working group, the LHC Higgs cross section working group, the NNPDF collaboration, the GFitter collaboration and the particle data group as well as Graeme Watt, who allowed me to use their material.

Furthermore, my thanks go to Ute Heuser, who organised the publication of this book at Springer, for her helpful advice on organisational matters.

Contents

Chapter 1
Introduction

The standard model of particle physics (SM) describes a large variety of results from accelerator experiments as well as cosmic ray observations. While the Large Hadron Collider (LHC) at CERN was build with the goals of discovering the remaining missing particle of the SM, the Higgs boson, as well as searching for physics beyond the SM, it serves also to gain a deeper understanding of the SM and its particles.

At the end of its initial running period more W and Z bosons have been produced in LHC collisions than any previous accelerator. The large number of electroweak bosons produced makes the LHC the preeminent laboratory for electroweak physics in current times.

While precision measurements of parameters of the SM are possible at the LHC, their accuracy is limited by the complicated initial and final states of the proton–proton collisions, as well as the large number of simultaneous collisions ("pile-up"). In many cases, previous measurements from lepton colliders are more precise with LHC results adding additional knowledge about the scale dependence of the parameters, as the high center-of-mass energy allows for a larger reach. The most notable exception is the W-boson mass. Already the currently most precise measurement of this quantity originates from a hadron collider and measurements at the LHC promise to improve this even further, though no result is public at this time.

The theoretically tractable electroweak bosons may serve as a clean probe into the QCD effects that form the proton into the complex object it is. The kinematic distributions of vector boson production provides information about the parton distribution functions (PDFs) of the proton. Measurements with electroweak bosons and heavy quarks provide important constraints on the flavor content, with the different couplings of the W and Z bosons complementing each other. Especially useful in this respect are separate measurements for positively and negatively charged W bosons, which are sensitive to the difference between quark- and antiquark-densities in the proton.

One of the most interesting features of the standard model is the mechanism of electroweak symmetry breaking, which imparts mass on the electroweak bosons via the Higgs mechanism. Naturally, studies of the symmetry breaking often involve electroweak bosons. This may take the form of searches for the Higgs boson, but

© Springer International Publishing Switzerland 2016
M.U. Mozer, *Electroweak Physics at the LHC*, Springer Tracts
in Modern Physics 267, DOI 10.1007/978-3-319-30381-9_1

also the study of multi-boson interactions, which are closely linked to the Higgs mechanism. While the analysis of Higgs boson decays to electroweak bosons can be used in this context, there is also the possibility to study the electroweak symmetry breaking by observing vector boson scattering, which is closely related to Higgs physics through interference effects. These processes are accessible for the first time at the LHC, as previous accelerators had neither the necessary luminosity nor center-of-mass energy.

Lastly, there is a wide variety of new physics models that would leave tell-tale signatures involving the electroweak bosons. Of particular interest in this respect are models that try to resolve the hierarchy problem through the introduction of extra dimensions.

In the following we'll give a brief description of the LHC, its two multi-purpose experiments (ATLAS and CMS) and the SM, followed by a description of the methods used to reconstruct W and Z bosons (collectively designated V, for vector-boson). The discussion of electroweak physics is divided into sections concerning electroweak bosons as probes into QCD, electroweak precision measurements, followed by descriptions of experimental results of Higgs physics, more exotic diboson resonances and non-resonant multi-boson studies.

1.1 Historical Overview

Initial theories of the weak interaction did not feature massive gauge bosons [1], but used four-fermion contact interactions. This simple description is very accurate at the low energies typical for the β-decays in which the weak interaction was first observed. However, it became clear that this model would not be renormalizable and lead to unitarity violations for interactions involving high energies, prompting the introduction of massive charged exchange bosons.

In the 1950s, major progress was made with the discovery that the weak interaction does not conserve parity [2, 3]. This result prompted major theoretical work, culminating in the proposal that the charged current vertex have the maximally parity violating $V - A$ structure [4]. In the following decade electroweak theory took approximately its current shape with the unification of the weak and electromagnetic interactions [5–7], which introduced the Z boson as neutral partner to the charged current exchanges. At the same time, the otherwise difficult to explain masses of the heavy gauge bosons were understood in terms of the Higgs mechanism [8–10]. This theory was put on solid ground with the proof that the theory is renormalizable [11], allowing for self-consistent predictions.

At this time, experimental evidence for the heavy vector bosons, especially the Z, was scant. The situation improved considerably with the discovery of weak neutral currents in the Gargamelle experiment at CERN [12, 13]. Direct observation of the gauge bosons had to wait until 1983, when the $Sp\bar{p}S$ proton anti-proton collider was taking data at beam energies high enough to produce real W and Z bosons [14–17].

These discoveries prompted major investments in facilities to study the electroweak bosons in detail, in the United States in the form of Stanford Linear Collider (SLC) and in Europe with the Large Electron Positron collider (LEP). Both were conceived to resonantly produce Z bosons in electron–positron collisions, allowing for the measurement of the structure and parameters of the electroweak interaction with unprecedented precision [18]. After producing about 17 million Z bosons, LEP was additionally operated at beam energies that allowed for the production of W boson pairs [19], contributing also to our knowledge of the W boson properties. Additional measurements of the W and Z boson were obtained at the Tevatron, most outstanding of which may be the measurement of the W boson mass [20], which required an enormous effort of the collaborations and has not yet been surpassed in precision.

With the start of the LHC in 2010 and the shutdown of the Tevatron in 2011, the center of experimental electroweak physics has shifted again to Europe. The results of the ongoing effort at the LHC will be the center of this work.

1.2 The Large Hadron Collider

The LHC is a proton–proton collider located at the French-Swiss border near Geneva, Switzerland. It reuses the 27 km diameter tunnel which originally housed the LEP electron positron accelerator (Fig. 1.1). In addition to protons, the LHC also collides heavy ions in dedicated data-taking periods. The protons are kept in orbit by 1200 dipole magnets, designed to produce a magnetic field of 8.3 T, corresponding to a beam energy of 7 TeV. The beams intersect at four points, each surrounded by a detector to study the resulting collisions. Two of these detectors, ATLAS and CMS, are general purpose experiments, designed to study a wide range of phenomena. The LHCb detector is a forward spectrometer, focused on the study of the decays of hadrons containing b quarks. The Alice experiment is optimized for the study of heavy ion collisions.

The design priorities of the LHC were TeV scale parton collisions at the highest possible luminosity, as theory studies [21] indicated that the motivating physics processes, such as the search for the Higgs boson or new physics beyond the SM may appear at the TeV scale and have low production cross sections. Originally planned to start operations in 1998 [22], the LHC neared completion in 2008. However, an incident during a power-test revealed shortcomings in the design and workmanship of the interconnects between dipole magnets. To avoid additional incidents, the LHC was started after repairs in 2010 with a beam energy of 3.5 TeV, which was increased to 4 TeV in 2012.

The LHC beam is structured into discrete proton "bunches", each containing upto 1.6×10^{11} protons with spacing as low 50 ns during the first running period ("Run I"). Taking into account larger gaps to facilitate beam injection and removal, a total of upto 1400 bunches were carried, for a total beam current of ~ 0.4 A . Under these conditions, the LHC achieved an instantaneous luminosity of

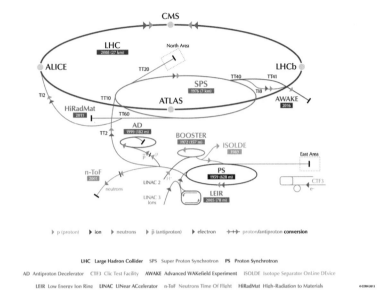

Fig. 1.1 Conceptual sketch of the CERN accelerator complex, showing the LHC and its pre-accelerators (Source: CERN.)

$0.65 \cdot 10^{34}$ cm^{-2}s^{-1}. However, these high numbers were not reached instantaneously. During 2010, only $\sim$0.045 fb^{-1} in integrated luminosity were collected, increasing to $\sim$5 fb^{-1} in 2011 and adding another $\sim$20 fb^{-1} in 2012.

During 2013 and 2014, the LHC operations were suspended for maintenance and a revision of the interconnects responsible for the reduced beam energy during the first running period. With the consolidated accelerator complex, beam energies of 6.5 TeV are foreseen for the next running period ("Run II"), starting in the summer of 2015.

1.2.1 The Compact Muon Solenoid

The Compact Muon Solenoid (CMS) detector [23] follows the common pattern of previous collider experiments in covering a large solid angle fraction around the interaction point with devices to measure the energy, momentum and particle species of outgoing collision products. The main feature of CMS is a 12.5 m long superconducting solenoid with a field strength of 3.8 T and an inner diameter of 6.3 m. Silicon track detectors for the measurement of the momenta of charged particles as well as calorimeters are located within the magnet, while muon chambers based on several different gaseous detector technologies are integrated into the flux return

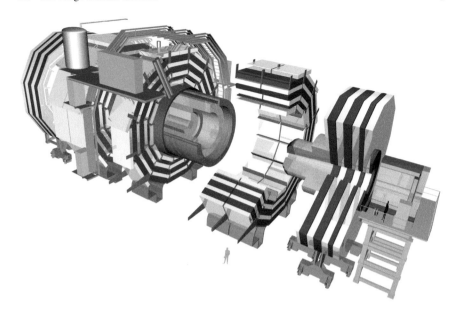

Fig. 1.2 The CMS detector showing the pixel detector (*brown*), tracker (*beige*), ECAL (*green*), HCAL (*yellow*), solenoid (*grey*), muon system (*white*) and flux return yoke (*red*) (Source: CMS/CERN.)

yoke (see Fig. 1.2). The CMS experiment uses a coordinate system centered on the nominal interaction point, such that the x-axis points towards the center of the LHC ring, the y-axis points upward and the z-axis is aligned with the beam-line to form a right-handed coordinate system. The azimuthal angle ϕ is measured from the x-axis in the $x - y$ plane and the polar angle θ from the z-axis. However, more commonly the pseudorapidity $\eta = -\ln\tan(\theta/2)$ is used instead.

The track detector comprises an inner part, built from silicon pixel modules, and an outer part, constructed with silicon strip detectors. Each of these parts are arranged into a barrel section with detectors parallel to the beam-pipe and an endcap section with modules perpendicular to the beam-pipe. The pixel detector close to the beam-pipe is optimized to provide the best resolution for the measurement of the point of origin of charged particles (the so called vertex), which is of particular importance in the identification of jets arising from the hadronization of heavy quarks. The pixel technology, with active regions of 100×150 µm does not only provide the necessary vertex measurement precision but also allows for a reasonably low detector occupancy even in the very dense particle flux close to the collision point, which aids in the identification of charged particle tracks without ambiguity. The pixel tracker consists of three layers in the barrel region and two endcap discs on each side, accounting for a total of 65 million pixels and covering a pseudorapidity range upto 2.5.

Further away from the interaction point, starting at 20 cm, occupancy is low enough to use the substantially more economical silicon strip technology. The strips are aligned parallel to the magnetic field in order to obtain maximum precision for the measurement of the charged particle momentum. Information about the track azimuthal angle is obtained from double-layers of detectors rotated by a stereo-angle of 40 mrad. The silicon tracker consists of a barrel-section as well as endcap discs on either side, both divided into an inner and outer sections. The tracker provides a position resolution between 23 and 53 μm perpendicular to the strip direction. The barrel and endcap layers are constructed such that a charged particle produced with $|\eta| < 2.4$ is guaranteed to pass at least 9 silicon layers, at least four of which provide two-dimensional information. While the silicon tracking detector of CMS has excellent track reconstruction efficiency and resolution, it comprises a substantial amount of material in front of the calorimeters: upto two radiation lengths at $|\eta| \approx 1.5$, complicating the reconstruction of electrons and photons.

The CMS experiment employs Kalman filter techniques [24] in order to reconstruct tracks in a computationally efficient manner even in the high occupancy events typical for hadron collisions under high pile-up conditions. Tracks are reconstructed in six iterations, treating the pixel- and strip detectors in a unified manner. Each iteration starts with a set of seeds, the first one being constructed from hits in the first three pixel layers using stringent quality and kinematic constraints. The seeds are used to initialize the Kalman filter, which incorporates further detector hits along the trajectory extrapolated from the previously included hits. The collected hits are finally fit together to estimate the charged particles kinematic quantities. The used hits are removed and the procedure repeats using seeds constructed from increasingly inclusive sets of tracking detector hits [25].

The electromagnetic calorimeter (ECAL) is located outside of the tracking detector. It is build as a homogeneous calorimeter using $PbWO_4$ crystals and divided in a barrel and endcap section, reaching pseudorapidities of $|\eta| < 3$. In contrast to the tracking detector, which is installed entirely within the solenoid, only the ECAL barrel is mounted directly to the magnet cryostat, while the ECAL endcaps are fixed to the detachable endcap sections of CMS. The gap between the barrel and endcap at $|\eta| \approx 1.5$ simultaneously serves to carry cable- and cooling infrastructure for the tracker and ECAL barrel to the outside, significantly reducing the ECAL performance in this region. The $PbWO_4$ was chosen for high density (8.3 g/cm^3), short radiation length (0.89 cm) and small Molière radius (2.2 cm). Additionally, the material has very fast scintillation decay time, which is important due to the LHCs small bunch spacing. ECAL crystals measure 22×22 mm on the face pointing towards the detector center, commensurate with the Molière radius and corresponding to a size of 0.0174×0.0174 in the $\eta - \phi$ plane. Crystals are 25.8 radiation lengths long, guaranteeing the full containment of all but the most energetic electromagnetic showers. The crystals are mounted pointing slightly off the nominal interaction vertex to avoid gaps aligned with the direction of particles emerging from the vertex. The ECAL reaches a resolution better than 2 % in the endcap and is mostly limited by the dynamic properties of the calorimeter: The $PbWO_4$ crystals lose transparency due to the irradiation on the time-scale of single data-taking runs and partially recover on similar timescales. To keep track of this effect, the calorimeter is equipped with a laser system to inject light pulses of known magnitude into each crystal and keep track of transparency losses.

The hadron calorimeter (HCAL) is constructed as a sampling calorimeter using brass as absorber and plastic scintillator as active material. It is segmented into barrel and endcap regions similarly to the ECAL, with the barrel covering an η range of $|\eta| < 1.4$ and the endcap reaching upto $|\eta| = 3$. In the forward direction, the HCAL is supplemented with a forward calorimeter (called HF) that extends the η coverage upto $|\eta| = 5.5$, which is especially useful to have a good acceptance for Vector-boson scattering processes. Due to the high flux of energetic particles in the forward direction, the HF is build using quartz fibers as active material. The quartz fibers do not scintillate, instead Čerenkov light is produced and gathered, as traditional scintillators will quickly degrade in the harsh radiation environment. The HCAL is segmented into towers of $\Delta\phi \times \Delta\eta = 0.087 \times 0.087$ in size, resulting in a five times lower granularity than the ECAL. Due to space constraints within the CMS solenoid, the HCAL has only an effective thickness of ~6 interaction lengths at $\eta = 0$, so that very energetic jets may not be fully contained in the calorimeter. An additional layer of scintillation counters is placed directly outside of the solenoid (corresponding to another two interaction lengths of material) to serve as "tail catcher". Overall, the HCAL achieves an energy resolution noticeably worse than the other detector components described above. To mitigate this issue and obtain competitive jet res-olutions, the CMS experiment exploits particle flow techniques described in more detail in Sect. 1.3.3.

The CMS solenoid is surrounded by an iron flux return yoke, which is inter-leaved with muon detectors. The muon detectors are built as gas ionization detectors in three different technologies: drift tubes are used in the barrel region and cathode strip chambers in the endcap to obtain muon track information. Additionally, resistive plate chambers (RPCs) are located in barrel and endcap to improve the muon trigger-ing capabilities of the experiment: The excellent time resolution of the resistive plate chambers is necessary to associate muon system signals to a particular collision, as the drift times in the gaseous tracking detectors is long compared to the average time between collisions. The barrel part of the muon system uses a combination of drift tubes and RPCs in four layers and is divided into five wheels, the central of which carries the central solenoid. The four other sections are not permanently installed around the coil but can be moved along the beam axis to allow access for main-tenance. The outer sections reach upto a pseudo-rapidity of $|\eta| = 1.1$. Overlapping with the barrel region, the muon endcaps start at $|\eta| = 0.9$ and extend upto $|\eta| = 2.4$. Three endcap discs with a total of four layers of cathode strip chambers and RPCs are installed on each side of the experiment. The muon endcaps serve as support for the calorimeter endcaps and can be moved similar to the outer barrel muon sections.

In addition to the subsystems described above, the CMS experiment contains a number of smaller sub-components which will not be discussed in detail here, such as a pre-shower detector in front of the ECAL endcaps and a very forward calorimeter for the study of diffractive physics, as well as a number of beam monitoring systems.

The CMS detector uses a two-stage trigger system: the so called L1 trigger uses purpose-built hardware to monitor the detector output and obtain a trigger decision within 3.2 μs at a rate of upto ~100 kHz. Due to the relatively long read-out latency of the tracking detector and the complexity of track reconstruction, the silicon tracker is not used in the L1 trigger. The decision is entirely based on information from the muon system and calorimeters. Acceptance by the L1 electronics triggers the full read-out of the detector. The data are sent to the surface, where they enter the second stage of the CMS trigger, the so called High Level Trigger (HLT). The HLT is constructed as a large farm of commodity computers (13,000 cores at the end of Run I), which run a somewhat reduced version of the CMS reconstruction software used for all data analysis. The system is designed to reach a decision within 200 ms for a total event rate of ~300 Hz. In addition to the full event data used in data analysis, there is a large number of collisions recorded with reduced detector information for calibration studies.

1.2.2   The ATLAS Detector

The ATLAS detector [26], shown in Fig. 1.3, is composed of the same basic components as the CMS detector. However, different technology choices were made for many of these components. The ATLAS detector uses complex set of magnet coils instead of the the simple solenoid of CMS: in addition to a solenoid surrounding the central tracking detectors, the ATLAS detector comprises a set of toroidal air core magnets in the outer layers of the detector, that allow the precise measurement

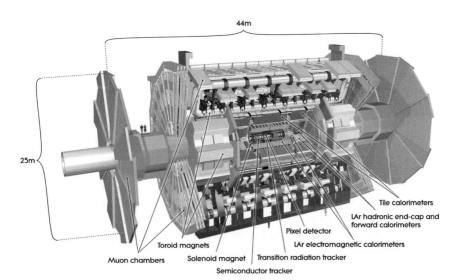

Fig. 1.3 The ATLAS detector showing the tracking detectors (*dark grey*), LAr calorimeters (*orange*), tile calorimeter (*grey*), magnets (*light grey*), muon system (*light blue*) (Source: ATLAS/CERN [26].)

of muons with little disturbance from multiple scattering processes. Overall, field strengths in the ATLAS detector (2 T for the solenoid, 1 T for the endcap toroids and 0.5 T for the barrel toroids) are lower than in the CMS detector, though the overall momentum resolution is similar as the ATLAS detector employs longer lever-arms.

At the center of the ATLAS detector, closest to the nominal interaction point is a silicon pixel detector composed of a three layer barrel section and three discs on either side, covering pseudorapidities upto $|\eta| < 2.5$. The pixels are markedly elongated, having dimensions of $50 \times 400 \, \mu m$, with the short dimension along the ϕ direction, in order to maximize the transverse momentum (p_T) resolution with a reasonable number of pixels. To achieve the best possible vertex resolution, the innermost pixel layer is located just 4.5 cm away from the nominal interaction vertex, just outside of the beam pipe.

Beyond the pixel detector, starting at a radius of 25 cm, a silicon strip detector takes over for similar reasons as discussed above for the CMS experiment. The pseudorapidity coverage is matched to the pixel detector, extending to $|\eta| < 2.5$. The ATLAS silicon tracker provides a resolution of 17 μm in the $R - \phi$ plane and is structured in 4 barrel layers and 9 endcap discs on either side. Each of the layers contain two sets of detectors at a stereo angle of 40 mrad in order to provide z-axis measurements. The comparatively low number of layers is warranted by the presence of another tracking detector outside of the silicon tracker: the transition radiation tracker (TRT), starting at radii of $\sim$55–60 cm. The TRT is based on drift tubes with a radius of 4 mm, interleaved with a plastic radiator material, which induces passing relativistic charged particles to emit transition radiation at γ-ray energies. The transition radiation is absorbed in the Xe and CO_2 based drift gas, leading to different specific ionization for particles of different mass. The effect is most pronounced for the lightest particles and serves to identify electrons in the ATLAS detector. Overall the TRT comprises 73 planes of tubes in the barrel region and 160 planes in the endcaps, reaching a similar pseudorapidity coverage of $|\eta| < 2$. The complete ATLAS tracking system contains upto two radiation lengths of material, peaking at $\eta \sim 1.6$. The whole tracking detector is surrounded by the solenoid, which contributes additional material in front of the calorimeters.

Track reconstruction in the ATLAS detector follows similar ideas as used in CMS, also employing Kalman filter techniques, but differs in many details, such as the number of iterations, definitions of track seeds and parameters. The TRT hits can be included in the tracks either as part of the initial track-finding procedure or alternatively, tracks can be reconstructed in the inner tracker, which then serve as seeds to the track-finding algorithm in the TRT. To reconstruct tracks that may be problematic to find using this approach, such as electrons with large energy loss or particle decays at large radii, there is the additional possibility to start track reconstruction at the TRT, extrapolating inwards, to recover some efficiency for these difficult cases [27, 28].

The electromagnetic calorimeter (ECAL) is built as a sampling calorimeter, using lead as absorber and liquid Ar (LAr) as active material. The absorber is shaped in an "accordion" geometry to avoid any cracks in the ϕ direction. The ionization signal is read out with very fine granularity, as much as 0.025×0.025 in the $\eta - \phi$ plane with three longitudinal segments. This fine granularity allows for an analysis of the shower shape in order to distinguish electrons and photons from hadronic interactions. The ECAL is segmented into a barrel and endcap region, with a transition at $|\eta| \sim 1.4$ and reaching a coverage upto $|\eta| < 3$. The calorimeter has a minimal depth of 22 (24) radiation lengths in the barrel (endcap) regions, respectively, fully containing the electromagnetic showers. The main advantage of the LAr technology is its stability and radiation hardness, allowing for a precise and stable calibration.

The ATLAS detector contains three separate hadron calorimeters, each optimized for the detector region it resides in and the corresponding radiation environment. In the barrel and outer endcap region, the hadron calorimeter is built with a steel absorber and plastic scintillator tiles (TileCal). The region behind the ECAL endcap uses LAr technology similar to the ECAL itself. The forward region ($3.2 < |\eta| < 4.9$, roughly equivalent to the CMS HF) is covered by the forward calorimeter (FCAL), also built in LAr technology. As the ATLAS calorimeters are not confined by the solenoid magnet, they are built to guarantee the containment of even the most energetic expected jets, adding upto ~ 10 interaction lengths in the central region and more in the forward region.

Beyond the calorimeters, the muon system is installed. The main toroidal magnet is instrumented by three stations of drift tubes, each comprising six to eight layers of drift tubes (DTs). Drift tubes are also employed in less irradiated parts of the endcap muon system. In the inner layers of the endcap muon system at large pseudorapidities ($2.0 < |\eta| < 2.7$), CSCs are used due to their tolerance of the harsh radiation environment in this region. The DTs and CSCs are supplemented with RPCs and thing gap chambers (TGCs), which provide fast readout for triggering. The high field strength of the toroidal magnets in combination with the large lever arm in the muon system allows for a very precise measurement of muon momenta without the use of the central tracking detector.

The trigger system of the ATLAS detector is structured in three layers. The first layer (L1), is built from custom hardware and uses only the muon system and calorimeters to reduce the data rate to ~ 100 kHz from the nominal interaction ratio of 20 MHz in the LHC Run I. In order to limit the depth of the electronic data buffers, the L1 decision has to be made within ~ 3 μs, precluding the use of tracking information. The second trigger level (L2) accesses all subdetectors in so called regions of interest, defined by the L1 logic. This step reduces the data rate to ~ 2 kHz in ~ 40 ms. The last step of the ATLAS trigger chain is a computer farm, similar to the CMS HLT, which reduces the final output rate to ~ 200 Hz, using the full detector data and event reconstruction.

Similar to the CMS experiment, the ATLAS detector contains a number of additional minor components, which are beyond the scope of this document. Overall it is remarkable to note, that the CMS and ATLAS experiments reach very similar performance, even though they select very different approaches to the challenges of data-taking in the harsh environment of high luminosity pp colliders.

1.3 Particle Reconstruction and Identification

The analysis are performed in terms of final state particles reconstructed from the electronic signals of the detectors. The correspondence is straight forward for the light leptons, but more complex for final state quarks and gluons, which are reconstructed as collimated jets of hadrons. Neutrinos may only be indirectly observed in the imbalance of the transverse momenta of the visible final state particles (E_T^{miss}). The following gives a short overview of the reconstruction of the particles most relevant for electroweak physics: electrons, muons, quarks and gluons as jets and neutrinos as E_T^{miss}.

1.3.1 Electrons

Electron reconstruction [29–31] has two major components in the ATLAS and CMS detectors: the collection of the electrons energy deposits in the ECAL ("clustering") and the reconstruction of the electrons track. Both are complicated by the fact that the track detectors contains between 0.4 and 2 radiation lengths of material, depending on the pseudorapidity, with additional material from the solenoid in the case of ATLAS. This large amount of material leads to a high probability for an electron to emit Bremsstrahlung radiation. In the clustering procedure this is taken into account by collecting energy deposits in a rectangular region that is wider in the ϕ direction than in the η direction, in order to collect the energy of the emitted photons. Tracker hits in the vicinity of the ECAL cluster are combined to a trajectory using the Gaussian sum filter [32], which allows for a change in track curvature for each traversed tracker layer, thus taking into account the energy loss from photons potentially radiated. In CMS the four-momentum of the electron is ultimately determined from the weighted average of the track and ECAL measurements, where the track measurement dominates for low electron momenta, while the ECAL information is most important for high momenta. In ATLAS, the energy is measured from the calorimeter alone, including corrections from a pre-sampling detector at the very front of the calorimeter to correct for losses in the material in front of the calorimeter, while the direction of is determined from the track fit.

Under optimal conditions both experiments achieve an electron energy resolution of about 0.5–1 % for high energy electrons in the central detector region, where there is little material in front of the calorimeters. The resolution degrades to 2–2.5 % when lower momenta are concerned and can reach as far as 10 % for electrons undergoing showers in the tracker material, as is common at rapidities of $\eta \sim 1.5$. The efficiency of successful electron reconstruction is of the order of 90–95 %, increasing with energy, and is limited by the chance to successfully match the track to the reconstructed cluster.

A major source of background for electrons are photons that undergo a pair-production process when traversing material of the tracker ("conversions").

This background is suppressed with several methods: electron tracks are required to contain hits in the inner tracking layers. Photon conversions in the inner layers are reduced by requirements on the compatibility of the electron track with the primary vertex.

As opposed to electrons emerging from the hard scattering process ("prompt electrons"), so called non-prompt electrons constitute the majority of the remaining backgrounds. They are mostly associated to QCD processes and originate from heavy quark decays or represent hadrons misidentified as electrons (often an ECAL cluster from a neutral pion decay combined with a close-by track of a charged hadron). These backgrounds are reduced by a suite of requirements collectively referred to as "electron id" [29–31]. The electron id uses a number of variables connected to the compatibility of the position of the ECAL cluster and the track impact point, quantities describing the distribution of energy within the ECAL cluster and the presence of energy in the HCAL directly behind the electrons ECAL cluster. In the ATLAS experiment, information from the TRT is used in addition.

To further suppress non-prompt electrons, isolation criteria are commonly employed. Electron candidates originating from hadrons are often seen in close proximity to further hadrons due to the dynamics of QCD shower development. In contrast, the leptons originating from the decay of electroweak bosons typically emerge without accompanying additional particles. This behavior is exploited by collecting detector deposits in a cone of radius $\Delta R = \sqrt{\Delta \eta^2 + \Delta \phi^2}$ around the electron and setting a threshold above which the electron candidate is rejected. Several different detector signatures have been used: separate isolation variables using tracks, ECAL- and HCAL deposits, or the sum of all three. Alternatively particles reconstructed with the particle flow algorithm (see below) may be used. Most commonly the transverse momenta of the deposits is added upto obtain the isolation variable ("absolute isolation") or divided by the transverse momentum of the electron candidate ("relative isolation").

The efficiency of the electron isolation is dependent on the pile-up as additional particles for pile-up interactions may randomly fall into the isolation cone. The isolation is thus computed using only tracks that originate from the same vertex as the electron candidate. Additionally, the expected contribution from pile-upto calorimetric variables are subtracted, either in an average subtraction dependent on the number of reconstructed vertices or with a pile-up density computed for each event [30], scaled to the area of the isolation cone.

In the case of electrons an additional complication arises from the propensity of the electrons to emit Bremsstrahlung photons when traversing the material of the track detector. If not all of these deposits are collected in the electron reconstruction, they may appear in the ECAL isolation or track isolation after conversion. In order to remove these deposits, the area over which the isolation is computed may exclude a narrow cone around the electron direction as well as a thin strip in ϕ.

The impact of these additional requirements on the overall selection efficiency varies, as the criteria are chosen to suit the individual analysis. In studies with high backgrounds, the efficiencies can be as low as 80 % overall, and even lower for forward electrons, where backgrounds are highest.

1.3.2   Muons

Muons are reconstructed using the gas detectors of the muon systems and the central tracking detector [33, 34]. Signals in the muon chambers are combined in a trajectory fit to form so called standalone muons. These are combined with tracks from the central tracker to form "global" (CMS) or "combined" (ATLAS) muons, which have improved kinematic resolution due to the high performance of the central tracking systems. If especially high efficiencies or muons with low momentum are important, it is possible to use a third reconstruction technique: "tracker muons" (CMS) and "segment-tagged muons" (ATLAS) are reconstructed from any track in the central tracking detector compatible with any deposit in the muon system. The improved efficiency is balanced by increased backgrounds, though. Especially energetic hadrons which may have showers that extend into the first layer of the muon system are increased compared to global/combined muons. The ATLAS collaboration additionally uses "calorimeter-tagged muons", a combination of a track in the inner detector and a calorimeter deposit compatible with a minimum ionizing particle. This category has the largest backgrounds but can cover regions where the muon system coverage is insufficient for the other types.

The momentum resolution for muons in the LHC detectors is on the order of 1–5 %. The resolution is best in the central detector region for low p_T muons. In contrast to the electrons, muon resolution degrades with increasing momentum, as the momentum measurement is driven by the track measurement, reaching $\sim$10 % for TeV muons. The efficiency to successfully reconstruct a muon is about 95 % for global/combined muons and rises to almost 100 % if tracker-/segment-tagged muons are included as well.

Selections on parameters related to the fits in the tracker, the muon-system and the combination are used to suppress badly reconstructed muon candidates and backgrounds. Requirements are based on the number of observed and expected hits in the detector as well as goodness of fit variables. Cosmic muons that travel through the detectors may be reconstructed as two outgoing muons originating from the point of closest approach to the beam-line. This type of background can be reduced by requiring the muons to be compatible with the primary vertex of the event.

For similar reasons as is the case for electrons, isolation variables are used to reject non-prompt muons. The general methods used are the same as for electrons, though detailed definitions may differ. The most prominent difference is connected to the higher mass of the muon: as muons are much less likely to emit Bremsstrahlung photons, it is not necessary to remove certain areas from the isolation cone as is the case for electrons.

1.3.3   Jets

Beyond the special cases of the electrons and muons, all CMS detector signals are reconstructed using the particle flow algorithm, which ultimately results in a set of reconstructed particles, classified as charged hadrons, neutral hadrons or photons [35, 36]. The algorithm combines charged particle tracks reconstructed in the central tracking detector with clusters from the ECAL and HCAL, weighting the information from each subdetector with the appropriate uncertainty. Thus, the strong magnetic field and excellent track resolution of the CMS detector minimize the impact of the comparatively poor resolution of the hadron calorimeter. In the ATLAS experiment, jets are formed from so called topological clusters, which represent topologically connected calorimeter cells with energies above the expected noise levels [37].

Jets are formed by clustering the complete set of particles or calorimeter clusters using one of several jet algorithms. The most common choice is the anti-k_T algorithm [38] with a radius parameter of R $= 0.5$ ("AK5 jets", used in CMS) or R $= 0.4$ (used in ATLAS). This algorithm is infrared- and collinear safe [39] and provides a comparatively stable jet area [40], which eases calibration in the busy environment of hadron collisions. The jet radius was chosen to balance uncertainties between corrections for particles outside of the jet area and foreign contributions from pile-up and the underlying event. For specialized studies involving jet substructure, large radius parameters and other jet algorithms (mostly the Cambridge–Aachen [41] and k_T [42] algorithms) may be used.

The jet algorithms used by the CMS and ATLAS collaborations fall into the class of sequential combination algorithms. They start from a set of particles or calorimeter clusters ("pseudojets"). A pairwise distance measure is computed and the two pseudojets with the smallest distance are combined into one new pseudojet, usually by adding the four-momenta. This procedure is iterated until all pseudojets are separated from each other by more than some cut-off distance.

Jet energy calibrations are extracted from the studies of the transverse momentum balance in di-jet events as well as events where a single photon or Z boson is produced recoiling against a jet [37, 43–45]. Additionally, contributions from pile-up particles are statistically subtracted event by event. In the CMS experiment, the average particle density outside of clustered jets is determined event by event for this purpose. Correction factors are then derived using this pile-up density and the jet area [46]. In the ATLAS experiment, pile-up corrections are derived from simulation and applied as function of the number of expected and observed simultaneous collisions.

In both experiments, the jet p_T resolution is about 10 % for 100 GeV jets, falling to ~5 % at very large p_T. In spite of the corrections for pile-up, the resolution of low p_T jets is still degraded for high pile-up data-taking periods. Compared to the electrons and muons discussed above, the jet reconstruction also suffers from significant scale uncertainties. In CMS and ATLAS these have been reduced in dedicated studies [37, 44] to below 1 % for central jets above 100 GeV, but uncertainties remain larger for jets at low p_T or in the forward regions of the detector.

Jets arising from the hadronization of heavy quarks (i.e. b and c quarks, as top quarks decay before they hadronize) are of special importance in the study of electroweak processes: hadronic Z decays have a much higher fraction of c and b quarks (~27 % together) than most QCD processes. Additionally, top quark decays, which constitute a significant background to studies with W bosons, have a ~100 % branching ratio to b quarks.

Heavy quark jets may be identified in the detector from several signatures absent in jets arising from the hadronization of light quarks or gluons:

- The weak decays and corresponding long lifetimes of heavy quark hadrons can give rise to tracks arising from a secondary vertex, measurably displaced from the collisions primary vertex.
- The decays of the comparatively high mass of heavy quark hadrons produces daughter particles with large transverse momentum relative to the jet axis.
- The weak decays of heavy quark hadrons contain high branching ratios to final states with leptons, which can be easily identified.

Selecting jets according to these criteria will greatly enrich the sample in b-quarks, and less so in c quarks, which are significantly lighter. The CMS and ATLAS experiments have used a number of different algorithms exploiting this information, ranging from a simple discriminant based on track impact parameters only to sophisticated multivariate discriminants using the complete track kinematics and secondary vertex information associated to a jet [47, 48].

1.3.4 Missing Transverse Energy

While neutrinos do not leave signatures in the detector, they can still be reasonably identified by the momentum they carry away undetected. As the total momentum is conserved, the sum of all detected momentum is equal in magnitude and opposite in direction to the sum of neutrino momenta. However, the CMS and ATLAS detectors have good coverage only in the transverse direction, while a large fraction of the final state momentum parallel to the beam axis leaves the experiment undetected in the beam pipe. Accordingly, only the sum of the transverse momenta of the neutrinos can be reconstructed by taking the negative vectorial sum of the transverse component of reconstructed deposits (E_T^{miss}).

In the analysis from the CMS collaboration presented here, E_T^{miss} is generally reconstructed from the PF particles associated to the primary vertex [49, 50], though other studies have used calorimeter deposits or tracks exclusively. The result is corrected for the energy scale corrections of jets included in the E_T^{miss}. In the ATLAS experiment, E_T^{miss} is computed using the calorimeter and muons (as described above), using object-specific calibrations for calorimeter deposits associated to a given type of object (i.e. electrons, photons, jets) [51]. The E_T^{miss} resolution is studied in $Z \rightarrow \ell\ell$ events, where the lepton signatures are artificially removed from the E_T^{miss} reconstruction.

The resolution that can be achieved by these methods is at the level of 10 GeV for $E_T^{miss} = 500$ GeV to 20 GeV for $E_T^{miss} = 1.5$ TeV. Even including all available corrections for pile-up effects, the E_T^{miss}-resolution still degrades with increased pile-up. In analysis that are particularly sensitive to the E_T^{miss}, the E_T^{miss}-vector is often split into components parallel and perpendicular to prominent recoil objects, as these two components may have markedly different resolutions.

Instead of the absolute value of E_T^{miss}, it is possible to compare the measured value of E_T^{miss} to it's uncertainty to obtain the E_T^{miss} significance. This variable is constructed as the ratio of the likelihoods of the hypothesis with E_T^{miss} equal to the value measured with the PF algorithm and the null hypothesis ($E_T^{miss} = 0$) [49]. While the E_T^{miss} significance cannot directly be used to reconstruct the kinematic properties of an event, it serves well in distinguishing events containing one or more neutrinos and events that do not contain neutrinos, where the observed E_T^{miss} arises from mismeasurements of particles in the detector.

References

1. E. Fermi, Versuch einer theorie der β-strahlen. Z. Phys. **88**, 161–177 (1934). doi:10.1007/BF01351864
2. T. Lee, C.-N. Yang, Question of parity conservation in weak interactions. Phys. Rev. **104**, 254–258 (1956). doi:10.1103/PhysRev.104.254
3. C. Wu et al., Experimental test of parity conservation in beta decay. Phys. Rev. **105**, 1413–1414 (1957). doi:10.1103/PhysRev.105.1413
4. R. Feynman, M. Gell-Mann, Theory of Fermi interaction. Phys. Rev. **109**, 193–198 (1958). doi:10.1103/PhysRev.109.193
5. S. Glashow, Partial symmetries of weak interactions. Nucl. Phys. **22**, 579–588 (1961). doi:10.1016/0029-5582(61)90469-2
6. A. Salam, J.C. Ward, Electromagnetic and weak interactions. Phys. Lett. **13**, 168–171 (1964). doi:10.1016/0031-9163(64)90711-5
7. S. Weinberg, A model of leptons. Phys. Rev. Lett. **19**, 1264–1266 (1967). doi:10.1103/PhysRevLett.19.1264
8. F. Englert, R. Brout, Broken symmetry and the mass of gauge vector mesons. Phys. Rev. Lett. **13**, 321–323 (1964). doi:10.1103/PhysRevLett.13.321
9. G. Guralnik, C. Hagen, T. Kibble, Global conservation laws and massless particles. Phys. Rev. Lett. **13**, 585–587 (1964). doi:10.1103/PhysRevLett.13.585
10. P.W. Higgs, Broken symmetries and the masses of gauge bosons. Phys. Rev. Lett. **13**, 508–509 (1964). doi:10.1103/PhysRevLett.13.508
11. G.'t Hooft, Renormalizable lagrangians for massive Yang-Mills fields. Nucl. Phys. B **35**, 167–188 (1971). doi:10.1016/0550-3213(71)90139-8
12. Gargamelle Neutrino Collaboration, Observation of neutrino like interactions without muon or electron in the gargamelle neutrino experiment. Phys. Lett. B **46**, 138–140 (1973). doi:10.1016/0370-2693(73)90499-1
13. F. Hasert et al., Search for elastic ν_μ electron scattering. Phys. Lett. B **46**, 121–124 (1973). doi:10.1016/0370-2693(73)90494-2
14. UA1 Collaboration, Experimental observation of isolated large transverse energy electrons with associated missing energy at $\sqrt{s} = 540$ GeV. Phys. Lett. B **122**, 103–116 (1983). doi:10.1016/0370-2693(83)91177-2
15. UA1 Collaboration, Experimental observation of lepton pairs of invariant mass around 95 GeV/c^2 at the CERN SPS collider. Phys. Lett. B **126**, 398–410 (1983). doi:10.1016/0370-2693(83)90188-0
16. UA2 Collaboration, Observation of single isolated electrons of high transverse momentum in events with missing transverse energy at the CERN anti-p p collider. Phys. Lett. B **122**, 476–485 (1983). doi:10.1016/0370-2693(83)91605-2
17. UA2 Collaboration, Evidence for Z$^0 \rightarrow e^+e^-$ at the CERN anti-p p collider. Phys. Lett. B **129**, 130–140 (1983). doi:10.1016/0370-2693(83)90744-X

18. The ALEPH, DELPHI, L3, OPAL, SLD Collaborations, the LEP Electroweak Working Group, the SLD Electroweak and Heavy Flavour Groups, Precision electroweak measurements on the Z resonance. Phys. Rept. **427**, 257 (2006). doi:10.1016/j.physrep.2005.12.006. arXiv:hep-ex/0509008

19. The ALEPH, DELPHI, L3, OPAL Collaborations, the LEP Electroweak Working Group, Electroweak measurements in electron-positron collisions at W-boson-pair energies at LEP. Phys. Rept. **532**, 119 (2013). doi:10.1016/j.physrep.2013.07.004. arXiv:1302.3415

20. CDF and D0 Collaboration, 2012 update of the combination of CDF and D0 results for the mass of the W Boson, arXiv:1204.0042

21. G. Jarlskog (ed.), *ECFA Large Hadron Collider Workshop* (Aachen, Germany, 4–9 Oct 1990) Proceedings. 2

22. G. Jarlskog (ed.), *ECFA Large Hadron Collider Workshop* (Aachen, Germany, 4–9 Oct 1990) Proceedings. 1

23. CMS Collaboration, The CMS experiment at the CERN LHC. JINST **3**, S08004 (2008). doi:10.1088/1748-0221/3/08/S08004

24. R. Frühwirth, Application of Kalman filtering to track and vertex fitting. Nucl. Instrum. Meth. A **262**, 444–450 (1987). doi:10.1016/0168-9002(87)90887-4

25. CMS Collaboration, Description and performance of track and primary-vertex reconstruction with the CMS tracker. JINST **9**(10), P10009 (2014). doi:10.1088/1748-0221/9/10/P10009. arXiv:1405.6569

26. ATLAS Collaboration, The ATLAS experiment at the CERN large hadron collider. JINST **3**, S08003 (2008). doi:10.1088/1748-0221/3/08/S08003

27. ATLAS Collaboration, Operation and performance of the ATLAS semiconductor tracker. JINST **9**, P08009 (2014). doi:10.1088/1748-0221/9/08/P08009. arXiv:1404.7473

28. T. Cornelissen et al., Concepts, design and implementation of the ATLAS new tracking (NEWT). Technical Report ATL-SOFT-PUB-2007-007. ATL-COM-SOFT-2007-002, CERN, Geneva, Mar 2007

29. CMS Collaboration, Electron reconstruction and identification at $\sqrt{s} = 7$ TeV. CMS Phys. Anal. Summary CMS-PAS-EGM-10-004 (2010)

30. CMS Collaboration, Performance of electron reconstruction and selection with the CMS detector in proton-proton collisions at $\sqrt{s} = 8$ TeV. JINST **10**(06), P06005 (2015). doi:10.1088/1748-0221/10/06/P06005. arXiv:1502.02701

31. ATLAS Collaboration, Electron reconstruction and identification efficiency measurements with the ATLAS detector using the 2011 LHC proton-proton collision data. Eur. Phys. J. C **74**(7), 2941 (2014). doi:10.1140/epjc/s10052-014-2941-0. arXiv:1404.2240

32. W. Adam et al., Reconstruction of electrons with the Gaussian sum filter in the CMS tracker at LHC. eConf **C0303241**, TULT009 (2003). doi:10.1088/0954-3899/31/9/N01. arXiv:physics/0306087

33. CMS Collaboration, Performance of CMS muon reconstruction in pp collision events at $\sqrt{s} = 7$ TeV. JINST **7**, P10002 (2012). doi:10.1088/1748-0221/7/10/P10002. arXiv:1206.4071

34. ATLAS Collaboration, Measurement of the muon reconstruction performance of the ATLAS detector using 2011 and 2012 LHC proton-proton collision data. Eur. Phys. J. C **74**(11), 3130 (2014). doi:10.1140/epjc/s10052-014-3130-x. arXiv:1407.3935

35. CMS Collaboration, Particle-flow event reconstruction in CMS and performance for jets, taus, and E_T^{miss}. CMS Phys. Anal. Summary CMS-PAS-PFT-09-001 (2009)

36. CMS Collaboration, Commissioning of the particle-flow event reconstruction with the first LHC collisions recorded in the CMS detector. CMS Phys. Anal. Summary CMS-PAS-PFT-10-001 (2010)

37. ATLAS Collaboration, Jet energy measurement and its systematic uncertainty in proton-proton collisions at $\sqrt{s} = 7$ TeV with the ATLAS detector. Eur. Phys. J. C **75**(1), 17 (2015). doi:10.1140/epjc/s10052-014-3190-y. arXiv:1406.0076

38. M. Cacciari, G.P. Salam, G. Soyez, The anti-k_t jet clustering algorithm. JHEP **04**, 063 (2008). doi:10.1088/1126-6708/2008/04/063. arXiv:0802.1189

39. G. Leder, Jet fractions in e+ e- annihilation. Nucl. Phys. B **497**, 334–344 (1997). doi:10.1016/ S0550-3213(97)00240-X. arXiv:hep-ph/9610552

40. M. Cacciari, G.P. Salam, G. Soyez, The catchment area of jets. JHEP **0804**, 005 (2008). doi:10. 1088/1126-6708/2008/04/005. arXiv:0802.1188

41. M. Wobisch, T. Wengler, Hadronization corrections to jet cross-sections in deep inelastic scattering (1998), arXiv:hep-ph/9907280

42. S.D. Ellis, D.E. Soper, Successive combination jet algorithm for hadron collisions. Phys. Rev. D **48**, 3160–3166 (1993). doi:10.1103/PhysRevD.48.3160. arXiv:hep-ph/9305266

43. CMS Collaboration, Determination of jet energy calibration and transverse momentum resolution in CMS. JINST **6**, P11002 (2011). doi:10.1088/1748-0221/6/11/P11002. arXiv:1107.4277

44. CMS Collaboration, Jet energy scale and resolution in the 8 TeV pp data. CMS Phys. Anal. Summary CMS-PAS-JME-13-004 (2015)

45. ATLAS Collaboration, Jet energy resolution in proton-proton collisions at $\sqrt{s} = 7$ TeV recorded in 2010 with the ATLAS detector. Eur. Phys. J. C **73**(3), 2306 (2013). doi:10.1140/ epjc/s10052-013-2306-0. arXiv:1210.6210

46. M. Cacciari, G.P. Salam, Pileup subtraction using jet areas. Phys. Lett. B **659**, 119–126 (2008). doi:10.1016/j.physletb.2007.09.077. arXiv:0707.1378

47. CMS Collaboration, Identification of b-quark jets with the CMS experiment. JINST **8**, P04013 (2013). doi:10.1088/1748-0221/8/04/P04013. arXiv:1211.4462

48. ATLAS Collaboration, Commissioning of the ATLAS high-performance b-tagging algorithms in the 7 TeV collision data. Technical Report ATLAS-CONF-2011-102, CERN, Geneva, July 2011

49. CMS Collaboration, Missing transverse energy performance of the CMS detector. JINST **6**, P09001 (2011). doi:10.1088/1748-0221/6/09/P09001. arXiv:1106.5048

50. CMS Collaboration, Performance of missing transverse momentum reconstruction algorithms in proton-proton collisions at $\sqrt{s} = 8$ with the CMS detector. CMS Phys. Anal. Summary CMS-PAS-JME-12-002 (2013)

51. ATLAS Collaboration, Performance of missing transverse momentum reconstruction in proton-proton collisions at 7 TeV with ATLAS. Eur. Phys. J. C **72**, 1844 (2012). doi:10.1140/epjc/ s10052-011-1844-6. arXiv:1108.5602

Chapter 2
Theory Overview

The SM is set up as a quantum field theory, using a Lagrangian formalism with gauge symmetry constraints to describe the matter particles and their interactions. The SM has been extraordinarily successful in describing the properties of matter and its interactions from subatomic to cosmological scales, and provides a unified view of the electromagnetic, weak and strong nuclear forces. Nevertheless, several open questions remain, such as the apparent presence of so called "dark matter" and "dark energy" in astronomical and cosmological surveys, the generation of neutrino masses and the observed matter-antimatter imbalance in the universe. A large variety of extensions of the SM have been devised to resolve these problems, though none of these have seen strong experimental confirmation yet. An overview of quantum field theories and their mathematical foundations may be found in the following textbooks: [1–5]. The description below largely follows Ref. [6].

2.1 The Standard Model Lagrangian

The Lagrangian of the SM $\mathcal{L}_{SM}$ may be split into several terms, each describing a different aspect of the underlying physics of the SM.

$$\mathcal{L}_{SM} = \mathcal{L}_{YM} + \mathcal{L}_{H} + \mathcal{L}_{ferm} + \mathcal{L}_{Yuk}. \tag{2.1}$$

The first term describes the gauge bosons and their interactions, as they arise from the gauge symmetries imposed on the Lagrangian and is accordingly denoted $\mathcal{L}_{YM}$ after Chen-Ning Yang and Robert Mills, who first analysed non-abelian gauge groups in depth [7]. A bare Yang-Mills theory requires massless gauge bosons contrary to observation. Accordingly, the second term $\mathcal{L}_H$ introduces the Higgs field, its self-interaction and interaction with the gauge bosons, which allow the gauge bosons to acquire mass in a gauge-invariant manner. The the third term subsumes the parts of the Lagrangian that describe the propagation of the matter fields and their interaction

© Springer International Publishing Switzerland 2016
M.U. Mozer, *Electroweak Physics at the LHC*, Springer Tracts
in Modern Physics 267, DOI 10.1007/978-3-319-30381-9_2

with the gauge bosons. Finally, $\mathcal{L}_{\text{Yuk}}$ describes the interaction of the matter fields with the Higgs boson, giving rise to the fermion masses through the Yukawa couplings.

$\mathcal{L}_{\text{YM}}$ can be written as:

$$\mathcal{L}_{\text{YM}} = -\frac{1}{4} W^i_{\mu\nu} W^{i,\mu\nu} - \frac{1}{4} B_{\mu\nu} B^{\mu\nu} - \frac{1}{4} G^a_{\mu\nu} G^{a,\mu\nu}, \tag{2.2}$$

where

$$W^i_{\mu\nu} = \partial_\mu W^i_\nu - \partial_\nu W^i_\mu - g\epsilon^{ijk} W^j_\mu W^k_\nu, \qquad i, j, k = 1, 2, 3, \tag{2.3}$$

$$B_{\mu\nu} = \partial_\mu B_\nu - \partial_\nu B_\mu, \tag{2.4}$$

$$G^a_{\mu\nu} = \partial_\mu G^a_\nu - \partial_\nu G^a_\mu - g_s f^{abc} G^b_\mu G^c_\nu, \qquad a, b, c = 1, \ldots, 8, \tag{2.5}$$

represent the field strength tensors associated to the different symmetries of the SM: $W^i_{\mu\nu}$ corresponds to the SU(2)$_{\text{I}}$ symmetry group of the weak isospin I^i_{w}, $B_{\mu\nu}$ to the U(1) symmetry of the weak hypercharge Y_{w} and $G^a_{\mu\nu}$ corresponding to the SU(3)$_{\text{c}}$ symmetry of the QCD color charge. ϵ^{ijk} and f^{abc} denote the structure constants of the SU(2) and SU(3) groups, following the conventions used in Ref. [8], respectively. g and g_s (as well as g' introduced below) denote the coupling constants for these interactions. For the B field, with its abelian U(1) symmetry, this term describes the free propagation of the field. For the W and G fields, with their non-abelian symmetries, additional terms arise, leading to interactions of the gauge fields among themselves. As the structure of these interactions is determined by the corresponding symmetry group, it is of interest to study multi-boson interactions to test the symmetry structure of the SM.

The interactions of the matter particles (i.e. fermions) with the gauge fields is described by

$$\mathcal{L}_{\text{ferm}} = i\overline{\Psi}_{\text{L}} \slashed{D} \Psi_{\text{L}} + i\overline{\psi}_{\ell_{\text{R}}} \slashed{D} \psi_{\ell_{\text{R}}} + i\overline{\Psi}_Q \slashed{D} \Psi_Q + i\overline{\psi}_{u_{\text{R}}} \slashed{D} \psi_{u_{\text{R}}} + i\overline{\psi}_{d_{\text{R}}} \slashed{D} \psi_{d_{\text{R}}}, \tag{2.6}$$

where Ψ_{L} represents left-handed lepton doublets of SU(2)$_{\text{I}}$, made up of the charged leptons and corresponding neutrinos, and Ψ_Q the equivalent doublets of up- and down-type quark pairs. $\psi_{f_{\text{R}}}$ denotes the corresponding right handed fermion singlets ($f = \ell, u, d$, where ℓ stands for charged leptons, u for up-type quarks, and d for down-type quarks), omitting the right-handed neutrinos, which have no interactions in the SM. As we will later see, the absence of right-handed neutrinos precludes mass generation for the neutrinos through interactions with the Higgs field. In the notation of Eq. 2.6, the interactions are hidden in the definition of the covariant derivative D:

$$D_\mu = \partial_\mu + ig I^i_{\text{w}} W^i_\mu + ig' Y_{\text{w}} B_\mu + ig_s T^a_{\text{c}} G^a_\mu, \tag{2.7}$$

where I^i_{w}, Y_{w} and T^a_{c} correspond to the generators of the respective gauge groups in the representation of the fermions they act on, as detailed in Ref. [8].

This description does not reduce trivially to the well established theory of quantum electrodynamics, where the fermions interact with the photon field A_μ in a manner that is parity-blind and proportional to $Q\overline{\psi} A \psi$. However, using the Gell-Mann–Nishijima relation for the electric charge $Q = I_w^3 + Y_w/2$, it is possible to recover the structure of quantum electrodynamics by constructing A_μ as a linear combination of the W_μ^3 and B_μ fields:

$$\begin{pmatrix} Z_\mu \\ A_\mu \end{pmatrix} = \begin{pmatrix} c_w & -s_w \\ s_w & c_w \end{pmatrix} \begin{pmatrix} W_\mu^3 \\ B_\mu \end{pmatrix} \tag{2.8}$$

Keeping the total normalizations constant, the linear combination is parameterized as a rotation by an angle θ_w, the so called weak mixing- or Weinberg angle. θ_w is determined by relating the unit charge e to the coupling constants g and g' as follows:

$$\cos\theta_w = c_w = \sqrt{1 - s_w^2} = \frac{g}{\sqrt{g^2 + g'^2}}, \qquad e = \frac{gg'}{\sqrt{g^2 + g'^2}}. \tag{2.9}$$

Through these relations, quantum electrondynamics is recovered as part of the SM, where in addition to the photon a second neutral gauge field, the Z boson arises. The remaining W_μ^1 adn W_μ^2 gauge fields have no definite electric charge and different linear combinations

$$W_\mu^\pm = (W_\mu^1 \mp i W_\mu^2)/\sqrt{2} \tag{2.10}$$

are chosen to represent the phyiscal fields with unit charge.

Considering only $\mathcal{L} = \mathcal{L}_{YM} + \mathcal{L}_{ferm}$, we arrive at a self-consistent theory, though all the involved particles, bosons as well as fermions, are massless. Masses cannot be easily introduced for either the bosons or the fermions, as the naive mass terms, $W_\mu^i W^{i,\,u}$ and $(\overline{\psi}_{f_L} \psi_{f_R} + \overline{\psi}_{f_R} \psi_{f_L})$ for the bosons and fermions, respectively, are not gauge invariant. For the fermions such a mass term would be a valid addition if left- and right-handed fermions behaved equivalently under $SU(2)_I \times SU(1)_Y$ transformations, but the absence of right-handed neutrinos in the SM spoils the symmetry. The generation of particle masses while preserving gauge invariance requires a more complex scheme.

2.1.1 Electroweak Symmetry Breaking and the Higgs Mechanism

The most commonly proposed mechanism to generate the masses of the SM particles is via the introduction of an additional symmetry which is spontaneously broken in the so called Higgs mechanism. It is introduced into the Lagrangian as

$$\mathcal{L}_H = (D_\mu \Phi)^\dagger (D^\mu \Phi) - V(\Phi), \tag{2.11}$$

where Φ is a complex scalar $SU(2)_1$ doublet $(\phi^+, \phi^0)^T$ with $Y_{w,\Phi} = 1$, leading to a positive electric charge for ϕ^+ and a neutral ϕ^0. The potential $V(\Phi)$ governs the self-interaction of the newly introduced field and may be freely chosen under the constraint that resulting Lagrangian is gauge invariant and renormalizable. The simplest form that fulfill these constrains and also allows for the generation of particle masses is:

$$V(\Phi) = -\mu^2(\Phi^\dagger\Phi) + \frac{\lambda}{4}(\Phi^\dagger\Phi)^2, \tag{2.12}$$

where μ^2 and λ are free real parameters. The condition $\lambda > 0$ guarantees a stable vacuum state. The sign on the first term is chosen to give the potential its characteristic "mexican hat" shape, which drives the spontaneous symmetry breaking: The ground state takes on a non-vanishing vacuum expectation value (vev) Φ_0. The vev is computed by minimizing $V(\Phi)$:

$$\Phi_0^\dagger\Phi_0 = \frac{v^2}{2}, \quad v = 2\sqrt{\frac{\mu^2}{\lambda}}. \tag{2.13}$$

The resulting ground state is not unique, but degenerate in three of its four dimensions. As the ground state remains symmetric under the unbroken $U(1)_{em}$ symmetry, the vev is only determined up to a complex phase, which we choose to give a real lower component to Φ_0. For the upper component, we choose a description that provides a vanishing value to obtain an electrically neutral vacuumm, i.e. $\Phi_0 = (0, v)^T$. The field Φ can thus be reparameterized in terms of perturbations around the vev:

$$\Phi = \begin{pmatrix} \phi^+ \\ \phi^0 = (v + H + i\chi)/\sqrt{2} \end{pmatrix}, \tag{2.14}$$

where H represents the real scalar Higgs field, which can be understood as a vacuum excitation and correspondingly carries the vacuum quantum numbers The additional fields ϕ^+ and χ are complex and real, respectively, and bear formal resemblance to Goldstone bosons. However, these three Goldstone-like modes are not physical, as a gauge transformation can always be found that will let them vanish. Using this so called "unitary gauge" Eq. 2.14 may be substituted together with the definition of the covariant derivative (Eq. 2.7) in Eq. 2.11, yielding

$$\mathcal{L}_{H,U\text{-gauge}} = \frac{1}{2}(\partial H)^2 + \frac{g^2}{4}(v + H)^2 W_\mu^+ W^{-,\mu} + \frac{g^2}{8c_w^2}(v + H)^2 Z_\mu Z^\mu \tag{2.15}$$

$$+ \frac{\mu^2}{2}(v + H)^2 - \frac{\lambda}{16}(v + H)^4, \tag{2.16}$$

using $I_{w,\Phi}^i = \sigma^i/2$, $Y_{w,\Phi} = 1$, $T_{c,\Phi}^a = 0$.

The results show the origin of mass for the electroweak gauge bosons: bilinear terms in the W and Z fields appear, proportional to the vev. This conserves the degrees of freedoms of the theory, the three Goldstone-like modes, which fell away in the unitary gauge, reappear as the additional degrees of freedom in the now massive gauge boson fields. A similar bilinear term in H is responsible for the mass of the Higgs boson itself. The masses may be expressed in terms of the previously defined parameters as:

$$M_{\mathrm{W}} = \frac{gv}{2}, \qquad M_{\mathrm{Z}} = \frac{M_{\mathrm{W}}}{c_{\mathrm{w}}}, \qquad M_{\mathrm{H}} = \sqrt{2\mu^2}. \qquad (2.17)$$

In addition to the masses, Eq. 2.16 also introduces interactions between the gauge bosons and the Higgs field as well as Higgs boson self interactions.

2.1.2 Fermion Masses

The generation of fermion masses can also be associated to the Higgs field, but proceeds by a fundamentally different mechanism. This is achieved by extending the SM Lagrangian by the so-called Yukawa term $\mathcal{L}_{\mathrm{Yuk}}$ that intermixes the fermions and the Higgs field in a gauge-invariant manner:

$$\mathcal{L}_{\mathrm{Yuk}} = -\overline{\Psi}_{\mathrm{L}} G_\ell \psi_{\ell_{\mathrm{R}}} \Phi - \overline{\Psi}_Q G_u \psi_{u_{\mathrm{R}}} \tilde{\Phi} - \overline{\Psi}_Q G_d \psi_{d_{\mathrm{R}}} \Phi + \mathrm{h.c.}, \qquad (2.18)$$

where "h.c." denotes hermitian conjugates and $\tilde{\Phi} = i\sigma^2 \Phi^* = ((\phi^0)^*, -\phi^-)^{\mathrm{T}}$ the charge-conjugate Higgs doublet with quantum numbers opposite to Φ. The G_f represent complex 3×3 matrices, which are free parameters of the theory. At first sight this appears to introduce a large number of free parameters into the SM. However, through appropriate field redefinitions a majority of these parameters may be eliminated.

The Yukawa term (Eq. 2.18) generates the fermion masses as it contains terms bilinear in the fermion fields. The fermion masses are encoded in the matrices G_f and are free parameters in contrast to the W and Z boson mass, which are strictly related to the weak couplings and the Higgs field parameters. Off-diagonal elements of the G_f induce oscillations between the fermion generations during free propagation. For the leptons, an appropriate basis can be chosen that diagonalizes G_ℓ, providing mass eigenstates of definite generation. This is not possible for the quarks, where the G_f for up- and down-type quarks cannot be diagonalized simultaneously. This leads to the mixing of quark generations in the weak interaction [9].

The absence of right-handed neutrinos in this setup prevents the generation of neutrino masses through this mechanism. However, since the initial work on the Higgs mechanism, neutrino oscillations have been discovered [10], indicating that neutrinos are massive, if light. The neutrino masses are commonly explained in terms of the so called see-saw mechanism [11, 12], though in the context of this work, neutrinos may effectively be treated as massless.

Using the mass eigenstates, each fermion couples to the Higgs boson with strength $y_f = M_f/v$. This coupling structure (i.e. purely scalar couplings proportional to the fermion mass) is a strong prediction of the SM Higgs mechanism and provides empirical means to distinguish it from alternative mass generation mechanisms, where other couplings structures and strengths may occur.

2.2   Predictions in Hadron Collisions

Perturbation theory can be used to compute the scattering matrix of processes involving the fundamental particles of the SM. However, color-charged particles like quarks and gluons are hidden from direct observation by the phenomenon of confinement [13]: at the LHC protons are accelerated and hadron jets detected in the experiments. These complex initial and final states are ultimately modeled to conform to our knowledge of strong interactions but must necessarily depend on ingredients derived from measurements to properly describe the low-energy, non-perturbative aspects of proton collisions.

The initial state protons may be envisioned in the parton model to consist of three valence quarks as well as a sea of virtual quarks and gluons. The composition of the proton is described by a parton desnsity function (PDF), that gives the probability to find a given parton carrying a momentum fraction x of the proton. Due to the non-perturbative effects prevalent in low energy QCD, the PDFs cannot be derived from first principles, but have to be extracted from data (see for example [14–16]). The PDFs are used to compute scattering cross sections, by integrating over all possible initial state momentum fractions:

$$\sigma_{pp \to X} = \sum_{i,j} \int dx_1 dx_2 \cdot pdf(x_1) pdf(x_2) \cdot \hat{\sigma}_{ij \to X}(x_1 P_1, x_2 P_2), \qquad (2.19)$$

where x_1 and x_2 denote the momentum fractions of the two interacting parton in their respective protons, P_1 and P_2 the proton momenta and $\hat{\sigma}_{ij \to X}$ the cross sections for two partons of type i and j to scatter to the final state X. The sum is taken over the types of partons in the proton, i, j. This naive approach suffers from two major issues: the cross section computed as described above is not stable against initial state QCD effects, as there is no well defined separation of scales between the PDFs and the hard matrix element. Additionally It is not clear whether such a simply factorized approach is possible at all.

The separation between the PDFs and matrix element can be made explicit by introducing a factorization scale Q^2, such that processes at an energy scale below Q^2 are implicitly included in the PDF definition, which gain a dependence on Q^2. The separating scale Q^2 is artificial, so that the theory can ultimately not depend on its value. This condition leads to a set of integro-differential equations (the so called DGLAP equations [17–19]), similar to the renormalization group equations, fixing the evolution in Q^2 of the PDF. This evolution describes the emission of gluons as well as gluon splitting starting from an initial parton in the proton. Though traditionally, the PDFs are defined for quarks and gluons, it is possible to also absorb initial state photon radiation into the PDFs, effectively resulting in a photon density of the proton [20]. The contribution of these photon induced processes is typically small for final states that can also be reached from quark- or gluon initial states. However, they may make up a significant fraction of the total cross section in processes involving electrically charged, but color-neutral particles, i.e. W bosons.

The applicability of the factorization approach shown in Eq. 2.19 can only be shown for some classes of all possible processes [21]. Generally, factorization is expected to hold in the limit of the production of very heavy or very high p_T particles. Nevertheless, experience shows that the factorization approach produces excellent predictions in the kinematic regime probed at the LHC.

Just as non-perturbative QCD effects complicate the initial state of hadron colliders, they are also responsible for complications in hadronic final states. Similar to the difference in scales between the scale of initial state proton mass and the hard scale of the scattering process, that is bridged by the DGLAP evolution, there is a difference in scales between the hard process and the scale of the final state hadron masses. The difference is treated analogously, producing gluon emissions off color-charged particles as well as gluons splitting in a process usually called parton shower. While a comprehensive treatment of the these emissions would be exceedingly complex, it turns out that color coherence effects suppress a large number of possible emission patterns, so that simple topologies dominate [22]. These simpler patterns, i.e. ordered in emission angle or transverse momentum, are the only ones used in the common simulation programs, greatly simplifying the computation.

Due to the confining nature of QCD, free color-charged particles are not observed, but rather color-neutral hadrons reach the detector. The transition between the partonic and hadronic regimes ("hadronization") is modeled empirically, with inspirations from the underlying theory. The simplest approach is based on so called fragmentation functions, which describe the probability to observe a given hadron carrying a certain momentum fraction of the parton. Current simulation programs use more sophisticated methods to model the hadronization process. The most prominent of these are the Lund string model [23] and the cluster fragmentation model [24]. In the Lund string model, the outgoing quarks are connected by strings, representing the confined color fields between the quarks. Potential energy stored in the string is used to iteratively create quark-antiquark pairs, forming color-neutral hadrons. In this picture gluons are envisioned as kinks in the connecting color strings. In contrast, the cluster model groups the final state partons into minimal color-neutral groups (the eponymous clusters), which are assumed to decay in a similar way as excited hadrons.

References

1. S. Weinberg, *The Quantum Theory of Fields* (Cambridge University Press, Cambridge, 1996)
2. M. Bohm, A. Denner, H. Joos, *Gauge Theories of the Strong and Electroweak Interaction* (Teubner, Stuttgart, 2001)

3. G.F. Sterman, *An Introduction to Quantum Field Theory* (Cambridge University Press, Cambridge, 1994)
4. R.K. Ellis, W.J. Stirling, B. Webber, QCD and collider physics. Camb. Monogr. Part. Phys. Nucl. Phys. Cosmol. **8**, 1 (1996)
5. J. Collins, Foundations of perturbative QCD. Camb. Monogr. Part. Phys. Nucl. Phys. Cosmol. **32**, 1 (2011)
6. T. Schörner-Sadenius (ed.), *The Large Hadron Collider: Harvest of Run I* (Springer, Berlin, 2015)
7. C.-N. Yang, R.L. Mills, Conservation of isotopic spin and isotopic gauge invariance. Phys. Rev. **96**, 191–195 (1954). doi:10.1103/PhysRev.96.191
8. S. Dittmaier, M. Schumacher, The Higgs boson in the standard model—from LEP to LHC: expectations, searches, and discovery of a candidate. Prog. Part. Nucl. Phys. **70**, 1–54 (2013). doi:10.1016/j.ppnp.2013.02.001. arXiv:1211.4828
9. M. Kobayashi, T. Maskawa, CP violation in the renormalizable theory of weak interaction. Prog. Theor. Phys. **49**, 652–657 (1973). doi:10.1143/PTP.49.652
10. K.A. Olive et al., (Particle Data Group), Review of particle physics. Chin. Phys. C **38**, 090001 (2014). doi:10.1088/1674-1137/38/9/090001
11. P. Minkowski, $\mu \to e\gamma$ at a rate of one out of 10^9 muon decays? Phys. Lett. B **67**, 421–428 (1977). doi:10.1016/0370-2693(77)90435-X
12. R.N. Mohapatra, G. Senjanovic, Neutrino mass and spontaneous parity violation. Phys. Rev. Lett. **44**, 912 (1980). doi:10.1103/PhysRevLett.44.912
13. G. 't Hooft, On the phase transition towards permanent quark confinement. Nucl. Phys. B **138**, 1 (1978). doi:10.1016/0550-3213(78)90153-0
14. R.D. Ball et al., Parton distributions with LHC data. Nucl. Phys. B **867**, 244–289 (2013). doi:10.1016/j.nuclphysb.2012.10.003. arXiv:1207.1303
15. A. Martin et al., Parton distributions for the LHC. Eur. Phys. J. C **63**, 189–285 (2009). doi:10.1140/epjc/s10052-009-1072-5. arXiv:0901.0002
16. H.-L. Lai et al., New parton distributions for collider physics. Phys. Rev. D **82**, 074024 (2010). doi:10.1103/PhysRevD.82.074024. arXiv:1007.2241
17. G. Altarelli, G. Parisi, Asymptotic freedom in parton language. Nucl. Phys. B **126**, 298 (1977). doi:10.1016/0550-3213(77)90384-4
18. Y.L. Dokshitzer, Calculation of the structure functions for deep inelastic scattering and e^+e^- annihilation by perturbation theory in quantum chromodynamics. Sov. Phys. JETP **46**, 641–653 (1977)
19. V. Gribov, L. Lipatov, Deep inelastic ep scattering in perturbation theory. Sov. J. Nucl. Phys. **15**, 438–450 (1972)
20. S. Dittmaier, M. Huber, Radiative corrections to the neutral-current Drell-Yan process in the Standard Model and its minimal supersymmetric extension. JHEP **1001**, 060 (2010). doi:10.1007/JHEP01(2010)060. arXiv:0911.2329
21. A. Mueller (ed.), *Perturbative Quantum Chromodynamics (Advanced Series on Directions in High Energy Physics)* (World Scientific, Singapore, 1989)
22. G. Marchesini, B. Webber, Monte Carlo simulation of general hard processes with coherent QCD radiation. Nucl. Phys. B **310**, 461 (1988). doi:10.1016/0550-3213(88)90089-2
23. B. Andersson et al., Parton fragmentation and string dynamics. Phys. Rept. **97**, 31–145 (1983). doi:10.1016/0370-1573(83)90080-7
24. B. Webber, A QCD model for jet fragmentation including soft gluon interference. Nucl. Phys. B **238**, 492 (1984). doi:10.1016/0550-3213(84)90333-X

Chapter 3
Experimental Signatures of EWK Bosons

Experimentally, the different decay channels of the electroweak bosons exercise every aspect of the detectors. Table 3.1 shows the branching ratios of the W and Z bosons: The leptonic decays (with the exception of the τ lepton) are particularly suitable where precise reconstruction and low backgrounds are of the highest importance. The hadronic decays suffer from large backgrounds and cannot be kinematically reconstructed as easily, but are commonly used to study processes with low cross section, due to the larger hadronic branching fractions of the bosons. The Z boson may also decay to a pair of neutrinos, which are not detected, so that such a decay can only be indirectly observed by studies of the recoil. This is only feasible if the recoil system is distinctive, such as the decay of a heavy resonance into a Z boson and another energetic particle.

3.1 Leptonic Decays

Electrons and muons can be reconstructed with excellent precision and purity in the ATLAS and CMS detectors and correspondingly, Z decays to electrons and muons are the preferred channel wherever the low branching fraction allows. The reconstruction of decays into τ leptons is complicated by the large number of different decay channels for the τ. The presence of at least one neutrino in each τ decay additionally degrades the kinematic reconstruction of these decays. For this reason, W and Z decays into τ leptons are usually not studied, though the $Z \rightarrow \tau\tau$ process has been measured [2, 3], mostly to study the performance of the τ reconstruction, as τ leptons play a major role in Higgs physics.

For the Z boson, typically two leptons of the same flavor and opposite charge are combined to obtain the four-momentum of the Z boson candidate. Selecting pairs with an invariant mass close to the nominal mass of the Z boson increases the contribution from Z bosons compared to virtual photons. Usually a mass window of 20–40 GeV around the Z mass is used, depending on the balance of sample purity and signal acceptance necessary for the analysis. In order to avoid backgrounds from

© Springer International Publishing Switzerland 2016
M.U. Mozer, *Electroweak Physics at the LHC*, Springer Tracts
in Modern Physics 267, DOI 10.1007/978-3-319-30381-9_3

Table 3.1 Branching Ratios for the W and Z boson [1]

W decay	Branching ratio (%)
$\rightarrow e\nu$	11
$\rightarrow \mu\nu$	11
$\rightarrow \tau\nu$	11
$\rightarrow q\bar{q}'$	67
Z decay	
$\rightarrow e^+e^-$	3.4
$\rightarrow \mu^+\mu^-$	3.4
$\rightarrow \tau^+\tau^-$	3.4
$\rightarrow \nu\bar{\nu}$	20
$\rightarrow q\bar{q}$	70
$\rightarrow b\bar{b}$	15

non-prompt leptons, reconstruction quality and isolation requirements as described in Sect. 1.3 area applied. As two leptons need to be misidentified simultaneously with the correct invariant mass, good purity can be achieved even with loose selection criteria.

Similarly, most W boson studies start with a single lepton, often with rather strict selection criteria, as the absence of a second lepton and mass selection dramatically increases the background from non-prompt leptons. The leptonic W decays include a neutrino, which escapes the experiment undetected and thus degrades the reconstruction of the W bosons kinematic quantities. If the final state contains only a single neutrino, the known mass of the W boson can be used in combination with the lepton kinematic quantities and the E_T^{miss} to reconstruct the four-momentum of the neutrino. As a replacement for the invariant mass of the lepton and neutrino, the transverse mass, M_T, is commonly used. M_T is defined as $\sqrt{2p_{T,\ell}E_T^{miss}(1-\cos\Delta\phi)}$, where $\Delta\phi$ is the angle in the transverse plane between the directions of the lepton and the E_T^{miss}, and $p_{T,\ell}$ is the transverse momentum of the lepton.

Using the known W boson mass and the assumption that the p_T of the neutrino equals E_T^{miss}, a quadratic equation for the longitudinal momentum of the neutrino can be derived with the solutions:

$$p_{z,\nu}^{\pm} = \frac{\mu p_{z,\ell}}{p_{T,\ell}^2} \pm \sqrt{\frac{\mu^2 p_{z,\ell}^2}{p_{T,\ell}^4} - \frac{E_\ell^2 p_{T,\nu}^2 - \mu^2}{p_{T,\ell}^2}}, \qquad (3.1)$$

where μ is defined as

$$\mu = \frac{M_W^2}{2} + p_{T,\ell}p_{T,\nu}\cos\Delta\phi. \qquad (3.2)$$

The ambiguity cannot be resolved with the information in the event alone and is complicated by the fact, that real solutions are not guaranteed. In case the measured transverse mass is larger than the W mass used to compute μ, no real solution is possible. This situation may easily arise due to mismeasurements, particularly of E_T^{miss}, as well as the natural width of the W boson. Reasonable performance can still be achieved by choosing the the solution with the smaller absolute value for the real case and the real part of the solution in the complex case.

3.2 Hadronic Decays

Hadronically decaying W and Z bosons can be equally well reconstructed by forming pairs of jets that result from the hadronization of the decay quarks. The resulting invariant mass peaks are rather broad due to the limited jet kinematic resolution of the detectors, so that the hadronic W and Z decays cannot be readily distinguished. Studies with hadronically decaying electroweak bosons suffer typically from large backgrounds, as the chance to produce two random jets from initial state QCD radiation with the invariant mass of the W or Z boson is substantial. In the case of the Z boson, a purer sample may be obtained by forming pairs from b-tagged jets, as the branching ratio of $Z \rightarrow b\bar{b}$ (15 %) is much larger than the chance to obtain b-jets in the hadronization of QCD processes or to misidentify two light jets. However, this approach will also reduce the signal cross section effectively by a factor of $\sim$5, so it is most suitable for processes with large cross sections, as an input in multivariate techniques or as a criterion for separating the dataset into high- and low-purity samples.

In most electroweak bosons produced at the LHC, the two decay quarks will hadronize into two separate jets. However, if the boson has a large momentum the two decay quarks will be separated only by a small angle, so that hadronization products from both quarks will overlap, effectively forming a single jet as shown

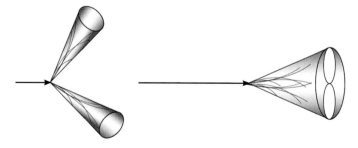

Fig. 3.1 Schematic view of hadronic V decays. For low momentum bosons (*left*) one distinct jet forms for each decay quark. For high momentum bosons (*right*), the decay products overlap, forming one single jet, though the hadronization products of each quark remain recognizable in the jet substructure

in Fig. 3.1. This final state, often called "boosted" or "merged" decay commands special treatment as it appears in particularly interesting processes such as high mass resonances decaying to dibosons or vector boson scattering at high energies.

Analysis using hadronic boson decays are hindered by the large cross section of QCD processes as well as the presence of pile-up, which may either add additional jets or add additional particles to primary jets, modifying their kinematic properties to more resemble boson decays. The contribution will also further broaden (and shift to higher masses) the invariant mass reconstructed for jet pairs originating from genuine bosons decays, further reducing the purity.

The LHC experiments have employed several different techniques to improve the reconstruction in Run I data [4, 5] and prepared several more for use in Run II [6–8], where conditions are expected to be even harsher. The reconstruction of the boson mass is improved by applying the appropriate jet energy corrections and removing jets that likely consist mostly of pile-up particles. The simplest techniques for this is to remove jets that have a large fraction of charged particles originate from a vertex other than the primary vertex. Additionally, it is possible to use the distribution of particles within the jet to discriminate jets arising from random associations of pile-up particles from jets formed in the hadronization of hard quarks or gluons [9]. The treatment of pile-up effects differs substantially between CMS and ATLAS due to the different nature of the jet reconstruction: In CMS, efforts are focused on using the track and vertex information inherent in the PF reconstruction, while ATLAS corrections are typically based on global event variables. Additionally, many ATLAS analysis use the momentum fraction of tracks originating at the primary vertex among all tracks geometrically linked to a jet to reject jets primarily consisting of pile-up particles.

The jets may also be corrected for pile-up on a more finely grained scale, trying to identify pile-up contributions constituent by constituent. This approach is greatly aided by PF reconstruction used by CMS, which allows to easily link vertex information obtained from the tracks to calorimetric information. The simplest choice here, is to remove all PF particles linked to vertices other than the primary vertex of the event from the jet reconstruction. While neutral pile-up particles remain, at least some fluctuations from the pile-up are removed, so that corrections can be smaller on average, reducing uncertainties. For Run II, a more sophisticated method is under consideration, the so called Pile-Up Per Particle Identification ("PUPPI") [10]. The algorithm uses the charged PF particles from the primary and other vertices to compute probabilities for the neutral PF particles to originate from pile-up or the primary collision, based on the geometric proximity to charged PF particles of known origin. Charged PF particles from pile-up vertices are discarded and the neutral PF particles four-vectors weighted according to their probability to be associated to the primary collision. The results are especially promising when applied to the reconstruction of boosted hadron decays, as discussed below, as the algorithm removes the detrimental effect that the pile-up has on jet substructure methods.

3.2.1 Decay to Dijets

The reconstruction of hadronically decaying electroweak bosons is conceptually straight forward: a pair of jets is chosen as hadronic decay candidate if the invariant mass of the pair is close to the boson mass. However, the mass resolution is poor compared to leptonic decays and there is a high chance for a random combination of jets not arising from boson decays to have the desired invariant mass. The latter issue is more severe in data with higher pile-up, as the additional pile-up particles promote jets into the signal region that would otherwise fail criteria on their p_T or add additional jets all by themselves. With the pile-up mitigation techniques discussed above, the effects can be minimized, but backgrounds are nevertheless high. To reduce the background further, several methods have been applied in the LHC experiments.

When the kinematic properties of the hadronically decaying Z are of interest, for example in the reconstruction of a resonance decaying to a pair of Z bosons, a kinematic fit can be employed to improve the the resolution of the kinematic Z reconstruction. For this fit, the p_T, η and ϕ of the two jets are varied to meet the nominal Z mass with minimal χ^2 of the variation. The procedure largely adjusts the jet p_T as the η and ϕ measurements have much smaller uncertainties than the p_T measurement. While this procedure improves, for example, the mass resolution in the reconstruction of diboson resonances, it complicates the use of events outside the di-jet mass selection as control sample: events in the such control samples are guaranteed to exhibit large χ^2 for the kinematic fit and have the jet kinematics adjusted by larger factors than the signal data.

Hadronic boson decays result in quark- anti-quark pairs, while most backgrounds are rich in gluon induced jets. This results in somewhat different distributions of particles within the jets, as the hadronization proceeds somewhat differently for the color singlet and octet states [11]. Experimentally, this can be used in the form of a likelihood discriminant, that determines if a jet is more quark- or gluon-like depending on such input variables as the multiplicities of charged and neutral PF particles and geometric moments of the constituents relative to the jet axis, as well as p_T moments. The discriminant is tuned on simulation, but is verified on data samples enriched in quark jets (photons produced in association with jets) and gluon jets (di-jet events), showing overall good agreement with data [12].

A more recent development is the so called jet pull variable [7, 13]. The variable tests for asymmetries in the two jets of a jet pair aligned with the connecting axis between the two jets. In hadronic boson decays these asymmetries are expected to arise due to the color connection of the two decay quarks, while no such asymmetry is expected for random pairs of jets.

3.2.2 Boosted Decays

If the boson has very high p_T, the two decay jets will be very close to each other in the detector system due to the strong Lorentz-boost. In the case of a heavy resonance X decaying to two bosons, this effect starts to occur when $M_X \geq \frac{M_V}{R/2}$, corresponding to $M_X \geq 360\,\mathrm{GeV}$ for the common jet radius of 0.5 and becomes dominating at about twice that mass. The single jet produced in such events is measurably different from jets produced in the hadronization of single quarks or gluons. The invariant mass formed by the combination of the four-vectors of all the jet-constituents (jet mass) is peaked at the vector boson mass while single quark and gluon jets present a steeply falling jet mass spectrum. Additionally, the two-pronged nature of the hadronic boson decays is imprinted on the geometric distribution of particles within the jet, while single quark and gluon jets have constituent distributions that are on average rotationally symmetric around the jet axis.

Both of these signatures can be used to distinguish hadronic decays of energetic vector bosons from other types of jets, allowing for a "V-tag" similar to the b-tag commonly used to identify heavy quark hadronization products. However, the jet mass in particular is affected by pile-up as well as particles from the underlying event. To limit this problem, algorithms have been studied, that clean a jet of constituents compatible with these background processes. As an added advantage, these so called jet-grooming techniques distinguish the patterns of jet mass generation observed in the parton shower (which gives jet mass to single quark and gluon jets) and boson decays (where the mass originates from two hard partons). As a result, the jet grooming algorithms sharpen the mass peak of hadronic boson decays and lower the average mass of single quark or gluon jets, resulting in better overall signal and background separation.

The algorithms are inspired by our understanding of the QCD parton shower, which underlies the formation of the jets. Initial work was focused on the quark jet masses (see for example Ref. [14]), which was studied at LEP to gain a deeper understanding of QCD effects at the edge of the perturbative regime. Since then, the calculations for quark and gluon jet mass observables have grown in sophistication (see Ref. [15] for a recent example), building the foundation of the background predictions used in analysis with boosted boson decays. Proposals to use substructure tools also arose in the LEP era [16], but were not directly applicable to this accelerator, due to a lacking beam energy. The variety of possible jet grooming and substructure techniques is large and a representative set will be discussed in more detail below. While the algorithms are largely inspired by the QCD shower evolution, it is not easily possible to map the finer points of QCD into an experimental algorithm. For this reason it is worthwhile to compare the results of the different algorithms in a full simulation with analytic calculations [17, 18], showing that some of them may show pathological behavior for extremely energetic jets or excessive dependence on the hadronization model used in the simulation.

In the LHC experiments, the following algorithms have been studied in detail [4, 6, 19, 20], visualized in Fig. 3.2:

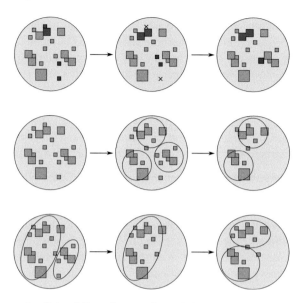

Fig. 3.2 Diagrams visualizing different jet grooming techniques. The *squares* represent calorimeter clusters or PF particles in η–ϕ space with an area proportional to their p_T. The algorithms start with the pseudojets contained in a large radius jets (*large gray circle*). See text for details. *Top* Pruning. The soft pseudojet in unbalanced (*red*) and large angle (*blue*) combinations are dropped (*crosses*) in favor of balanced combinations while reclustering the jet. *Middle* Trimming and Filtering. The jet is reclustered with a smaller jet radius and only subjets above a given p_T fraction (*Trimming*) or p_T rank (*Filtering*) are kept. *Bottom* Softdrop. The jet clustering sequence is undone step by step, discarding subjets that are soft compared to the combination

Trimming The jet-trimming algorithm [21] proceeds as follows: The constituents of the initial jet are reclustered with a jet algorithm using a smaller radius parameter R than the original algorithm. Constituents are removed if they are not part of one of the new small radius jets that fulfill the condition $p_{\mathrm{T}i} > f_{\mathrm{cut}} \cdot \Lambda$, where $p_{\mathrm{T}i}$ is the transverse momentum of a small radius jet, f_{cut} a tuneable parameter of the algorithm and Λ a suitable hard scale, typically the transverse momentum of the initial large radius jet.

Filtering The filtering procedure [22] proceeds similar to the trimming algorithm, but instead of retaining subjets with a given momentum fraction, a given number of subjets are retained.

Pruning The pruning algorithm [23] re-clusters the constituents of a given jet with the CA algorithm [24]. In each iteration of the CA algorithm two four-vectors i and j are combined if they are neither (i) separated by a large angle, nor (ii) are p_T^i or p_T^j small compared to the p_T of their combination. More exactly, (i) can be expressed as $\Delta R_{ij} > m^{\mathrm{orig}}/p_\mathrm{T}^{\mathrm{orig}}$ and (ii) as $\min(p_\mathrm{T}^i, p_\mathrm{T}^j)/\tilde{p}_\mathrm{T} < 0.1$, with $\tilde{p}_\mathrm{T}$ the transverse momentum of the combination of i and j, m^{orig} the mass and $p_\mathrm{T}^{\mathrm{orig}}$ the transverse momentum of the original jet.

Softdrop The softdrop algorithm [25] is a generalization of the previously used mass-drop algorithm [22]. The algorithm starts with a CA jet, iteratively undoing the clustering sequence. The jet is split into the two pseudojets it is composed of, with momenta p_{T1} and p_{T2} and geometric distance of Δ_{R12}. The softer of the two subjets is removed if $\frac{\min(p_{T1}, p_{T2})}{(p_{T1}+p_{T2})} < z_{\text{cut}} \cdot (\frac{\Delta_{R12}}{R})^{\beta}$, with R the radius parameter of the original jet and tuneable parameters. The procedure is repeated for the surviving subjets until it arrives at the initial particles which cannot be split anymore. Compared to the other algorithms discussed here, the soft-drop algorithm is infrared safe and thus amenable to use in analytic calculations.

The pruning algorithm has been extensively used by the CMS collaboration during Run I, where it showed excellent performance [4]. However, the algorithms effectiveness depends noticeably on the pile-up conditions, so that the CMS collaboration has reevaluated several techniques for Run II, focusing on the performance under the high pile-up conditions expected during Run II. Figure 3.3 shows the mass response for a variety of grooming algorithms in combination with several pile-up mitigation techniques. The most prominent improvement in resolution is associated to the introduction of any pile-up suppression. Beyond that, the different grooming algorithms perform similarly well in improving the core of the resolution function. However, the pruning algorithm and to a lesser extent the softdrop algorithm show considerable

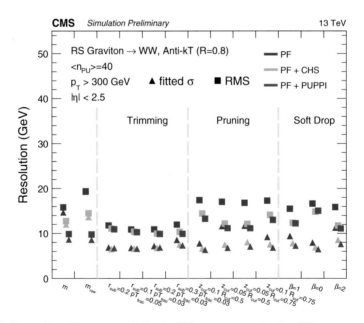

Fig. 3.3 Comparison of jet mass resolution for W jets reconstructed using different grooming and pile-up suppression algorithm. The resolution is evaluated from the RMS (*squares*) and σ (*triangles*) from a Gaussian fit to the full resolution distribution. The first two bins report the resolution of the ungroomed mass and of the raw mass (i.e. ungroomed and without any pile-up corrections) (Adapted from Ref. [6].)

tails in the resolution distribution as can be seen from the large difference between the σ from a Gaussian fit to the resolution and the RMS.

In addition to the improvements in jet mass reconstruction gained from jet grooming techniques, it is possible to recover information about the two boson decay quarks by studying the jet substructure [4, 7, 19, 20, 26]. The large number of proposed algorithms precludes a detailed discussion of all of them, instead we will provide an overview of algorithms used in LHC studies, with a more detailed review of a selected example.

splitting scale The splitting scale [27] corresponds to the distance value of the last recombination step of a jet clustered with the k_T algorithm.

momentum balance The momentum balance [22] is the splitting scale, as described above, normalized to the jet mass.

N-subjettiness The so called N-subjettiness variable [28] determines the compatibility of a given jet with N subjets, by computing the average normalized distance of the constituents to the closest of N subjet axis.

mass drop The change in jet mass in a given step of the jet clustering sequence can not only be used to groom the jet as described above, but large changes in jet mass are also indicative of subjet structure.

jet width The first moment of the distribution of p_T weighted distances (i.e. ΔR) of jet constituents to the jet axis discriminates between widely spread and centrally focused jets [29].

planar flow The planar flow is computed from the momentum correlation matrix of the constituents within a jet [30], reaching values close to one for jet with their constituents mostly aligned with the jet axis.

energy correlation functions In a somewhat more general approach energy correlation functions [31] use angular and energy correlation functions to identify a possible multi-prong structure without explicitly finding subjets.

qjets volatility When clustering the jet according to a recombination algorithm as described in Sect. 1.3.3, instead of combining the two closest constituents according to a given distance metric, the distance metric can be used as a combination chance, leading to a probabilistic clustering algorithm. Repeating the probabilistic clustering many times provides a set of different combination patterns, the distributions of which can be used to isolate sub-jet structures [32].

jet pull (subjets) The jet pull variables described above in the context of Vector boson decays into dijets can also be applied to subjets found within the boosted decays.

center-of-mass energy flow Energy flow variables such as the thrust, sphericity and
   acoplanarity [33, 34], originally devised for the study of the partonic structure in
   low energy collisions, can be used to distinguish boosted decays if applied to the
   jet constituents in the jet center-of-mass frame [35].

In addition to selection criteria based on any of these variables, they may also be
used in multivariate discriminants.

   While the members of the ATLAS experiment have tailored boosted jet identification techniques to each analysis separately, CMS has mostly standardized on a
single basic set of selection criteria for V-tagging in Run I based on the pruned mass
and the N-subjettiness. While this approach may not be optimal for every analysis
it provides synergies between analysis when determining efficiencies, misidentification rates and correction factors. Due to it's widespread use in CMS analysis with
boosted bosons, we will discuss the N-subjettiness [28], τ_N, in more detail.

   The N-subjettiness, τ_N, is used to measure the compatibility of a given jet with
having N subjets. To compute this variable, the constituents of the ungroomed jet
are re-clustered with the k_T algorithm, until N composite objects ("subjets") remain
in the iterative combination procedure of the k_T algorithm. The N-subjettiness, τ_N is
then defined as:

$$\tau_N = \frac{1}{d_0} \sum_k p_{T,k} \min(\Delta R_{1,k}, \Delta R_{2,k}, \dots, \Delta R_{N,k}), \qquad (3.3)$$

where the index k runs over the jet constituents, $\Delta R_{n,k}$ measures the geometric
distance between the k-th constituent and the n-th subjet axis and the normalization
factor d_0 is computed as $d_0 = \sum_k p_{T,k} R_0$, with R_0 denoting the radius parameter
of the original jet. Large values of τ_N indicate that a jet is not compatible with N
subjets, while small values show compatibility. In practice, the ratio between the
2-subjettiness and 1-subjettiness, $\tau_{21} = \tau_2/\tau_1$, is used as discriminating variable.

   To assess the performance of these techniques, a suitable sample of boosted boson
decays is necessary. While it is difficult to isolate a pure sample of hadronic Z decays,
such a sample can be obtained for W decays by studying semi-leptonic decays of $t\bar{t}$
production. Events are selected with a single lepton, E_T^{miss} and at least one b-tagged
jet to identify a top quark. Jets opposite to the leptonic top quark are enriched in
hadronic W decays as the single-lepton requirement suppresses fully leptonic top-
quark decays and the $t\bar{t}$ production cross section is much larger than the one for the
production of single top quarks. Figure 3.4 shows the resulting distributions of the
pruned mass and τ_{21} compared to simulation. While the mass distribution is very well
described by simulation, there are visible discrepancies for τ_{21}, so that appropriate
correction factors have to be applied when the efficiencies of τ_{21} selections are derived
from simulation.

   Similar to the di-jet case, b-tags can also be used in boosted boson decays to
enrich the sample in boosted hadronic Z decays. The ATLAS and CMS experiment
pursue different approaches in the use of this so called subjet b-tagging. At CMS,
the subjet b-tagging is performed with the same algorithms as the tagging of jets

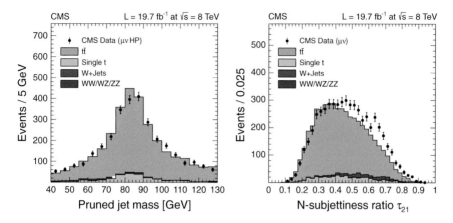

Fig. 3.4 Jet mass (*left*) and τ_{21} (*right*) distributions for high p_T jets in events with a lepton, a b-jet and E_T^{miss}, tagging semi-leptonic $t\bar{t}$ decays. The leptonic top candidate is required to be in the opposite hemisphere as the jet entering the figure (Adapted from Ref. [36].)

arsing from the hadronization of a single b quark [37], as described in Sect. 1.3.3. The efficiencies and mis-identification rates are extracted from simulation. However, compared to b-jets from single b quark hadronization, these values cannot be easily verified in data, so that sizable uncertainties have to be applied, based on comparisons in simulation and data with ordinary b-jets. Uncertainties become particularly large when the opening angle between the two subjets in the boosted final state is small, as the association of a given secondary vertex to a given subjet may be ambiguous. The techniques plays a relatively minor role in studies with boosted Z bosons, due to the comparatively low $Z \to b\bar{b}$ branching ratio, but is heavily used in the analysis of boosted top quark and Higgs boson decays, which are much more likely to contain b quarks. The ATLAS experiment, on the other hand, has reworked the subjet b-tagging from the ground up [38]. While this approach allows for optimal performance, it is more time and work intensive, leading to delays in publication.

The power of these methods is clearly demonstrated in Ref. [26]: Using a combination of several of the variables discussed above, the ATLAS researchers are able to measure the cross section for hadronically decaying W and Z bosons at high transverse momenta. While the background and corresponding uncertainties remain substantial, the result nevertheless shows the effectiveness of the substructure approach.

References

1. K.A. Olive et al., (Particle Data Group), Review of particle physics. Chin. Phys. C **38**, 090001 (2014). doi:10.1088/1674-1137/38/9/090001
2. CMS Collaboration, Measurement of the inclusive Z cross section via decays to tau pairs in *pp* collisions at $\sqrt{s} = 7$ TeV. JHEP **1108**, 117 (2011). doi:10.1007/JHEP08(2011)117. arXiv:1104.1617

3. ATLAS Collaboration, Measurement of the Z to tau tau cross section with the ATLAS detector. Phys. Rev. D **84**, 112006 (2011). doi:10.1103/PhysRevD.84.112006. arXiv:1108.2016
4. CMS Collaboration, Identification techniques for highly boosted W bosons that decay into hadrons. JHEP **1412**, 017 (2014). doi:10.1007/JHEP12(2014)017. arXiv:1410.4227
5. ATLAS Collaboration, Pile-up subtraction and suppression for jets in ATLAS. Technical Report ATLAS-CONF-2013-083, CERN, Geneva (2013)
6. CMS Collaboration, Study of pileup removal algorithms for jets. CMS Phys. Anal. Summary CMS-PAS-JME-14-001 (2014)
7. CMS Collaboration, V tagging observables and correlations. CMS Phys. Anal. Summary CMS-PAS-JME-14-002 (2014)
8. ATLAS Collaboration, Tagging and suppression of pileup jets with the ATLAS detector. Technical Report ATLAS-CONF-2014-018, CERN, Geneva (2014)
9. CMS Collaboration, Pileup jet identification. CMS Phys. Anal. Summary CMS-PAS-JME-13-005 (2013)
10. D. Bertolini et al., Pileup per particle identification. JHEP **1410**, 59 (2014). doi:10.1007/JHEP10(2014)059. arXiv:1407.6013
11. ALEPH Collaboration, Studies of quantum chromodynamics with the ALEPH detector. Phys. Rept. **294**, 1–165 (1998). doi:10.1016/S0370-1573(97)00045-8
12. CMS Collaboration, Performance of quark/gluon discrimination in 8 TeV pp data at CMS. CMS Phys. Anal. Summary CMS-PAS-JME-13-002 (2013)
13. J. Gallicchio, M.D. Schwartz, Seeing in color: jet superstructure. Phys. Rev. Lett. **105**, 022001 (2010). doi:10.1103/PhysRevLett.105.022001. arXiv:1001.5027
14. S. Catani, G. Turnock, B. Webber, Heavy jet mass distribution in e^+e^- annihilation. Phys. Lett. B **272**, 368–372 (1991). doi:10.1016/0370-2693(91)91845-M
15. M. Dasgupta et al., On jet mass distributions in Z+jet and dijet processes at the LHC. JHEP **1210**, 126 (2012). doi:10.1007/JHEP10(2012)126. arXiv:1207.1640
16. M.H. Seymour, Searches for new particles using cone and cluster jet algorithms: a comparative study. Z. Phys. C **62**, 127–138 (1994). doi:10.1007/BF01559532
17. M. Dasgupta et al., Towards an understanding of jet substructure. JHEP **1309**, 029 (2013). doi:10.1007/JHEP09(2013)029. arXiv:1307.0007
18. M. Dasgupta et al., Jet substructure with analytical methods. Eur. Phys. J. C **73**(11), 2623 (2013). doi:10.1140/epjc/s10052-013-2623-3. arXiv:1307.0013
19. ATLAS Collaboration, Performance of jet substructure techniques for large-R jets in proton-proton collisions at $\sqrt{s}$ = 7 TeV using the ATLAS detector. JHEP **1309**, 076 (2013). doi:10.1007/JHEP09(2013)076. arXiv:1306.4945
20. ATLAS Collaboration, Performance of boosted W Boson identification with the ATLAS detector. Technical Report ATL-PHYS-PUB-2014-004, CERN, Geneva (2014)
21. D. Krohn, J. Thaler, L.-T. Wang, Jet trimming. JHEP **1002**, 084 (2010). doi:10.1007/JHEP02(2010)084. arXiv:0912.1342
22. J.M. Butterworth et al., Jet substructure as a new Higgs search channel at the LHC. Phys. Rev. Lett. **100**, 242001 (2008). doi:10.1103/PhysRevLett.100.242001. arXiv:0802.2470
23. S.D. Ellis, C.K. Vermilion, J.R. Walsh, Recombination algorithms and jet substructure: pruning as a tool for heavy particle searches. Phys. Rev. D **81**, 094023 (2010). doi:10.1103/PhysRevD.81.094023. arXiv:0912.0033
24. M. Wobisch, T. Wengler, Hadronization corrections to jet cross-sections in deep inelastic scattering (1998), arXiv:hep-ph/9907280
25. A.J. Larkoski et al., Soft drop. JHEP **1405**, 146 (2014). doi:10.1007/JHEP05(2014)146. arXiv:1402.2657
26. ATLAS Collaboration, Measurement of the cross-section of high transverse momentum vector bosons reconstructed as single jets and studies of jet substructure in pp collisions at $\sqrt{s}$ = 7 TeV with the ATLAS detector. New J. Phys. **16**(11), 113013 (2014). doi:10.1088/1367-2630/16/11/113013. arXiv:1407.0800
27. J. Butterworth, B. Cox, J.R. Forshaw, WW scattering at the CERN LHC. Phys. Rev. D **65**, 096014 (2002). doi:10.1103/PhysRevD.65.096014. arXiv:hep-ph/0201098

28. J. Thaler, K. Van Tilburg, Identifying boosted objects with N-subjettiness. JHEP **03**, 015 (2011). doi:10.1007/JHEP03(2011)015. arXiv:1011.2268
29. ATLAS Collaboration, ATLAS measurements of the properties of jets for boosted particle searches. Phys. Rev. D **86**, 072006 (2012). doi:10.1103/PhysRevD.86.072006. arXiv:1206.5369
30. L.G. Almeida et al., Substructure of high-p_T Jets at the LHC. Phys. Rev. D **79**, 074017 (2009). doi:10.1103/PhysRevD.79.074017. arXiv:0807.0234
31. A.J. Larkoski, G.P. Salam, J. Thaler, Energy correlation functions for jet substructure. JHEP **1306**, 108 (2013). doi:10.1007/JHEP06(2013)108. arXiv:1305.0007
32. S.D. Ellis et al., Qjets: a non-deterministic approach to tree-based jet substructure. Phys. Rev. Lett. **108**, 182003 (2012). doi:10.1103/PhysRevLett.108.182003. arXiv:1201.1914
33. E. Farhi, A QCD test for jets. Phys. Rev. Lett. **39**, 1587–1588 (1977). doi:10.1103/PhysRevLett. 39.1587
34. J. Bjorken, S.J. Brodsky, Statistical model for electron-positron annihilation into hadrons. Phys. Rev. D **1**, 1416–1420 (1970). doi:10.1103/PhysRevD.1.1416
35. C. Chen, New approach to identifying boosted hadronically-decaying particle using jet substructure in its center-of-mass frame. Phys. Rev. D **85**, 034007 (2012). doi:10.1103/PhysRevD. 85.034007. arXiv:1112.2567
36. CMS Collaboration, Search for massive resonances decaying into pairs of boosted bosons in semi-leptonic final states at $\sqrt{s} = 8$ TeV. JHEP **1408**, 174 (2014). doi:10.1007/ JHEP08(2014)174. arXiv:1405.3447
37. CMS Collaboration, Performance of b tagging at $\sqrt{s} = 8$ TeV in multijet, ttbar and boosted topology events. CMS Phys. Anal. Summary CMS-PAS-BTV-13-001 (2013)
38. ATLAS Collaboration, Flavor tagging with track jets in boosted topologies with the ATLAS detector. Technical Report ATL-PHYS-PUB-2014-013, CERN, Geneva (2014)

Chapter 4
EWK Bosons and QCD

Many of the parameters of the electroweak interaction are known to high precision from previous experiments, particularly the LEP experiments and SLD [1, 2]. At the LHC there are natural limitations to the obtainable accuracy for such measurements due to the uncertainties inherent in the composite initial state of pp collisions. However, instead of having electroweak measurements limited by initial state QCD uncertainties, it is possible to revert the situation by assuming the electroweak parameters from previous experiments and using weak boson production to constrain the proton structure. In order to obtain accurate results, these studies focus on leptonic boson decays. The following will present a review of the possible ways electroweak bosons can be used to gain a deeper understanding of QCD and discuss a representative example in more detail.

4.1 Inclusive Production

Possibly the most basic measurement is the determination of the inclusive production cross sections of electroweak bosons. The couplings of the electroweak bosons have been measured to great accuracy at previous experiments [3] and the matrix element computed very precisely [4–9], so that the proton PDFs are the major uncertainty in the predictions of electroweak boson production cross sections. This, on the other hand, indicates that measurements of the electroweak boson cross sections may be used to constrain the PDFs. Figure 4.1 compares predictions of electroweak boson production using different PDFs, before the inclusion of LHC data. Note that in addition to the PDF set itself, the cross sections are correlated with the strong coupling α_S through the evolution of the parton densities and higher order corrections, complicating the analysis.

Experimentally, the measurements proceed by the reconstruction of weak boson candidates as described in Sect. 3.1, subtraction of background, corrections for

© Springer International Publishing Switzerland 2016
M.U. Mozer, *Electroweak Physics at the LHC*, Springer Tracts
in Modern Physics 267, DOI 10.1007/978-3-319-30381-9_4

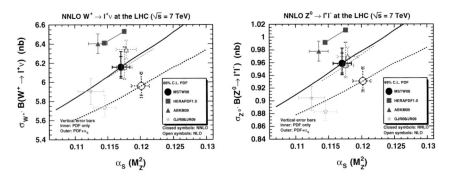

Fig. 4.1 W$^+$ (*left*) and Z (*right*) production cross sections in NNLO and NLO using different parton densities as function of α_S (Adapted from Ref. [10].)

reconstruction efficiency and normalization to the integrated luminosity of the dataset [11–13]. In inclusive electroweak boson production studies, the number of observed bosons is generally so large that statistical uncertainties are smaller than systematic uncertainties. Major systematic uncertainties for the inclusive cross section measurements are uncertainties in the lepton reconstruction for the Z production cross section and the E_T^{miss} reconstruction for the W production. Additionally, the uncertainty in the determination of the integrated luminosity contributes significantly. Cross sections are measured corrected for efficiency effects within the nominal acceptance ("fiducial cross section") as well as extrapolated to the full phase space ("total cross section"). Many of the most important systematic uncertainties cancel completely (e.g. luminosity) or partially (e.g. lepton reconstruction efficiencies) when the ratio of W and Z production cross sections are studied. Such a ratio measurement is quite sensitive to the proton PDFs due to the different flavor compositions of the initial states of W and Z production.

4.2 Differential Measurements

More information about the proton PDFs may be gained from differential measurements. Of particular interest is here the rapidity, y, of the produced boson as it is directly related to the momentum fractions of the initial state partons via the relation:

$$x_{1,2} = M_{\ell\ell}/s \cdot \exp(\pm y), \qquad (4.1)$$

with $x_{1,2}$ the momentum fractions of the two initial state partons, s the center of mass energy of the pp system and $M_{\ell\ell}$ the invariant mass of the outgoing boson decay products. In simple s-channel processes like inclusive W or Z production (see Fig. 4.2) the scale of the process is given as $Q^2 = x_1 \cdot x_2 \cdot s = M_{\ell\ell}$, so that measurements of the differential cross section as function of y and $M_{\ell\ell}$ directly scan the underlying PDFs [14, 15].

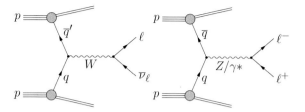

Fig. 4.2 Diagrams showing the Drell Yan process in pp collisions

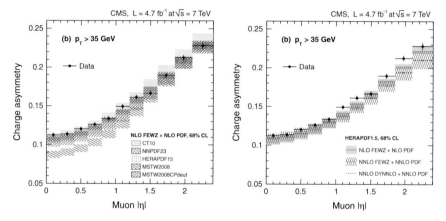

Fig. 4.3 Muon charge asymmetry in W boson decays in the CMS experiment compared to predictions of several PDFs in NLO (*left*) and NNLO (*right*) (Adapted from Ref. [16].)

Separate measurements of W^+ and W^- production allow the assignment of x_1 and x_2 to quarks and antiquarks, so that the differences in quark and anti-quark densities can be directly measured [16–18]. The major difficulty of such a measurement arises from the fact that the momentum component along the beam axis of the outgoing neutrino is not directly measured, so that the rapidity of the W cannot be directly reconstructed. The pseudo-rapidity of the outgoing lepton is well measured and correlated to the rapidity of the W boson and may serve as a proxy in such a measurement (see Fig. 4.3), at the cost of additional systematic uncertainties related to the prediction of the lepton pseudo-rapidity distribution for a given W rapidity distribution.

Of additional interest is the transverse momentum distribution in Drell-Yan events, which is zero in leading order. However, initial state radiation causes the boson p_T to take finite values in reality. The exact p_T spectrum is difficult to compute as the initial state emissions can be soft enough to be not easily accessible in perturbation theory. This theoretical uncertainty is one of the limiting factors in measurements of the W boson mass, as discussed in Chap. 5, and direct measurements of this effect in Z boson production [19–21] may help reduce this uncertainty.

In order to reduce the uncertainties in these measurements associated to the energy or momentum scales of the leptons, it is possible to test the calculations using variables only constructed using the angles of the outgoing leptons, as the opening angle between the leptons in the transverse plain is related to the boson p_T. The method was pioneered by the D0 collaboration at the Tevatron [22] and is currently employed by the ATLAS experiment [23].

The inclusion of the differential electroweak cross sections into parton density fits is attractive because computations at NNLO are available [7, 24], allowing the data to be included in fits of NNLO precision. The situation is different for example for multi-jet cross sections at hadron colliders, where no such computation exists and the corresponding data are either excluded [25] or treated in approximation [26]. For the W and Z production processes, electroweak corrections have been computed as well. However, the inclusion of these corrections are hindered by the fact that a consistent treatment is only possible if the photon content of the proton is included in the fit, which is not usually done. The NNPDF collaboration has at least partially included these corrections by separating QED corrections from the full electroweak corrections and excluding data points were the corrections are large, i.e. at high $M_{\ell\ell}$ and y.

4.3   EWK Bosons in Association with Jets

The properties of the underlying boson production mechanisms may be probed by measuring the production of vector bosons in association with jets. If the additional jets are studied for heavy quark content, the results provide insight into the heavy quark densities of the proton [27–30].

Processes in which the boson appears accompanied by light jets are ultimately sensitive to the strong coupling constant α_S. At leading order, the cross section to produce a boson in association with n jets is suppressed with respect to the cross section with $n - 1$ jets by a constant factor proportional to α_S [31, 32], the so called Behrends-Giele scaling. While the relation is not strictly true at higher orders, it is still approximately valid [33, 34]. It is important to verify this scaling experimentally, as many signatures of new physics, especially in supersymmetric theories, are known to exhibit final states with leptons, E_T^{miss} and a large number of jets. Estimating the expected background to such signatures from Behrends-Giele scaling may provide a valuable cross check to other background estimates for such signatures.

Predictions for this process are hindered by the large number of outgoing partons. Tree-level calculations have now been successfully automated [35] up large numbers of final state particles. When simulated event samples are to be obtained by combining calculations for different numbers of final state jets, care needs to be taken in order to avoid double counting that would otherwise be induced by the parton shower adding additional emissions that are already accounted for in the matrix elements of the next higher jet number computation [36, 37]. Considerable effort is required to compute the cross section of V+jet production in full NLO for larger numbers of jets. Current calculations reach up to five jets [38], a number that has rapidly increased over the last years. If simulated events are to be derived from such a computation, the matching procedure to avoid double counting becomes more complicated to take into account also the effect of the virtual corrections [35, 39] and correspondingly few of the available computational tools allow for the generation of simulated events. Thus, another goal of the experimental studies in this channel is to verify whether the computationally much less complex LO computations reasonably describe the observations to decide whether they may serve as adequate background models in the search for exotic new physics.

In the following we will discuss the analysis of electroweak boson production in association with jets from the ATLAS experiment [40, 41] in some detail, as it serves to demonstrate a range of methods and techniques widely used in other studies of electroweak physics. Similar measurements have been performed by the CMS experiment [42, 43] with comparable results.

4.3.1 Selection Outline

In the Z analysis, events with two leptons of the opposite charge and the same flavor are considered. The leptons are required to have $p_T > 20$ GeV and to be located in the detector region where lepton identification is highly efficient, i.e. $|\eta| < 2.4$ for muons and $|\eta| < 2.47$ for electrons with the additional exclusion of the endcap-barrel transition region $1.37 < |\eta| < 1.52$. In addition to the purely kinematic requirements, the leptons must pass a number of criteria designed to reject misidentified and non-prompt leptons, as discussed in Sect. 1.3. Events are considered as Z candidates if the dilepton invariant mass fulfills $66 < M_{\ell\ell} < 116$ GeV.

The corresponding W analysis selects events with single electrons or muons in the same pseudorapidity range as the Z analysis. The W analysis is based on events recorded on the decision of a single-lepton trigger, which employs higher p_T thresholds than the lepton-pair trigger used for the Z analysis and accordingly, the p_T threshold in the analysis is also higher: $p_T > 25$ GeV. Additionally, tighter lepton identification criteria are applied than in the Z analysis. A significant amount of E_T^{miss} is required ($E_T^{miss} > 25$ GeV), and the transverse mass of the lepton and E_T^{miss} system has to be larger than 40 GeV. These two conditions are highly correlated and serve to suppress QCD induced background with non-prompt leptons.

Jets are required to have $p_{T,j} > 30$ GeV and $|y| < 4.4$, uniformly for the W and Z analysis, easing comparison. To reduce the effect of pile-up the techniques described in Sect. 3.2 are employed.

4.3.2 Signal Extraction

Events are classified into categories based on the number of jets observed in addition
to the boson. The number of signal events is determined by subtracting the expected
background and correcting for the acceptance and efficiency of the analysis. The
lepton identification and isolation efficiencies of the simulation are corrected to
match the ones observed in the data. To compute these correction factors, the tag-
and-probe method is used. Two lepton candidates close to the Z mass are selected,
one well identified and isolated, the other only loosely identified. This sample is
mostly composed of real leptons, due to the large Z boson production cross section
compared to backgrounds with non-prompt leptons under these selection conditions.
The number of loosely identified leptons passing/failing the nominal identification
criteria can then be used to estimate the lepton identification efficiency, taking into
account the small remaining background. Efficiency studies are performed separately
in the different n-jet bins, as especially the efficiency of the isolation requirements
falls significantly with the number of additional jets in the event. In addition to the
lepton uncertainties, the effects of the jet energy scale and pile-up subtraction are
studied in detail.

In the Z channel, background is dominated by $t\bar{t}$ and QCD multi-jet processes,
with some contributions from electroweak diboson production. Overall backgrounds
are very low (<1 % in the one-jet category), due to the unique $M_{\ell\ell}$ signature of the
process. Figure 4.4 shows a representative di-lepton mass spectrum (electron channel,
Z + 1 jet category).

Fig. 4.4 Invariant mass
distribution of electron pairs
compared to simulation
(Adapted from Ref. [40].)

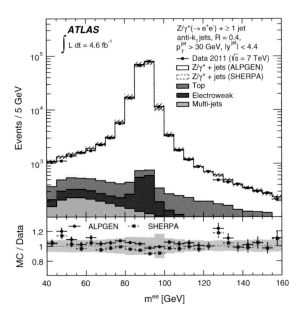

For the W bosons, background is overall much higher than for the Z and consists of mainly two components: a QCD induced component, where the lepton is the result of a heavy quark decay or arises from a misidentified meson and a component involving top-quark decays. Both backgrounds are determined from studies of suitable control regions. The two backgrounds exhibit very different M_T and E_T^{miss} spectra: the QCD background falls monotonously, in contrast, the $t\bar{t}$ background has a peak that coincides with the peak of the signal spectrum. The QCD background is determined by fitting simulation-derived templates to the low part of the E_T^{miss}, which is excluded from the nominal analysis. The $t\bar{t}$ background is determined in a similar template fit using the acoplanarity [44] spectrum of a sample enriched in $t\bar{t}$ production by a b-tag requirement.

The treatment of the $t\bar{t}$ background is markedly different from the corresponding CMS analysis [43], where this background is suppressed by a veto on b-tagged jets. Due to this extraction method, W+b-jet events are classified as background. This produces a reduced systematic uncertainty in the background estimate at the cost of increased uncertainties when extrapolating to the whole cross section.

4.3.3 Results

The raw spectrum of signal yield as function of jet multiplicity is converted to the resulting measured cross-section using an unfolding technique that accounts simultaneously for efficiency and migration between multiplicity bins. All unfolding methods are based on the same general idea: Simulation is used to prepare a so called migration matrix, which describes the probability of an event of given kinematic properties to be reconstructed with another set of observable properties. The observed data distribution may then be multiplied with the inverted migration matrix to obtain the physical distribution, corrected for the effects of detector efficiencies and resolution. However, this naive approach suffers from numerical instability: finite statistical uncertainty in the migration matrix can lead to unstable results in the matrix inversion. Additionally, the direct inversion will often enhance statistical fluctuations in the observed data, simultaneously introducing strong correlations between different points in the distribution, making the result hard to interpret by visual inspection.

Various more advanced unfolding techniques [45] have been devised to combat these issues by replacing the inverted migration matrix with an approximation that simultaneously reduces the problems discussed above. In the analysis discussed here, the d'Agostini iterative procedure [46] is used, while the corresponding CMS study uses the Singular Value Decomposition (SVD) [47]. Figure 4.5 shows the resulting distribution for the number of jets in the $Z \to ee$ channel. The jet energy uncertainty is the dominating systematic uncertainty, increasing with the number of jets. This increase can be canceled by measuring the ratio of cross sections of boson production in association with n jets divided by the number with $n-1$ jets, i.e. the Behrends-Giele scaling parameter.

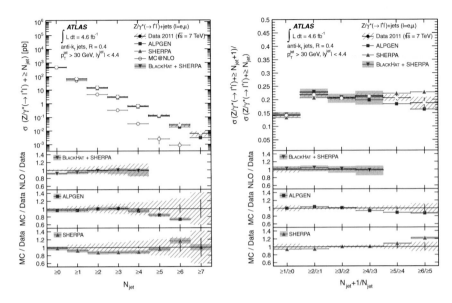

Fig. 4.5 Cross section of Z+jet production as function of the number of jets (*left*) and ratio of cross sections with jet numbers differing by one (*right*). Both results are compared to a number of theoretical predictions (Adapted from Ref. [40].)

The unfolding procedure has the advantage of correcting bin migration effects, but introduces correlated uncertainties between the bins, which makes it more difficult to perform a fit of the Behrends-Giele scaling to the resulting cross section or cross section ratio. To avoid this issue, the CMS collaboration has additionally studied the scaling directly [48].

The parameters of the scaling are instead determined with a forward-smearing approach. The cross section dependence on the jet multiplicity is parameterized as a linear function, with the constant term describing the pure scaling behavior and the slope parameterizing deviations from constant scaling. This function is folded with the migration matrix described above and used as normalization constraint in a simultaneous extended maximum likelihood fit to the different multiplicity bins of the invariant mass spectra. Figure 4.6 shows the corresponding fit result for both Z channels, indicating that the scaling behavior is consistent with expectations within the uncertainties.

4.3.4 Results with Heavy Quarks

Additionally, it is possible to use heavy quark tagging techniques as described in Sect. 1.3.3 to study the production of electroweak bosons in association with heavy jets. These processes give insight into the heavy quark densities in the proton, but are

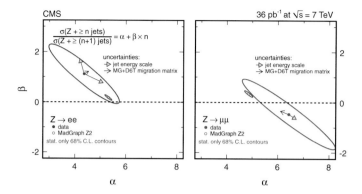

Fig. 4.6 Results of the fits to the Behrends-Giele parameters in Z+jets events. *Left* electron channel; *Right* muon channel (Adapted from Ref. [48].)

also of interest for their contribution to the backgrounds in searches for SUSY and similar exotic scenarios. Predictions involving b-quarks are particularly challenging, as the mass of the b-quark is heavier than the proton, but also much lighter than the typical hard scales present in LHC collisions, such as the b-quark transverse momentum. The available calculations simplify this in two different approaches: In the 4-flavor scheme, the b-qark is treated as massive and arises from pair creation in the hard process. In the 5-flavor scheme the b-quark is treated as massless and contributes directly to the proton PDFs.

The LHC experiments have studied the production of Z bosons in association with b quarks in detail [28, 30, 49]. The studies follow the approach discussed above to select leptonically decaying Z bosons, but differ significantly in their identification of b quarks. While Refs. [28, 30] employ b-tagged jets, Ref. [49] relies solely on the presence of secondary vertices identified with the tracking detector. The analysis agree that calculations in the 4-flavor scheme provide a somewhat more accurate description of the kinematic distributions of the b-qarks, while the 5-flavor scheme is more successful in describing the total cross section, especially for events with only one observed b quark.

The production of W bosons in association with b jets is dominated by processes involving the top quark and are performed mostly in the scope of measurements concerning the properties and dynamics of top-quark production. More directly connected to the proton PDFs is the production of W bosons in association with a single charm quark [27, 29], which is directly sensitive to the strange quark density of the proton. Leptonically decaying W bosons are reconstructed as described above. The charm quarks are identified by reconstructing $D^{\pm}$ and $D^{*\pm}$ meson decays, as well as semi-leptonic decays. Subtracting the measured distributions where the W boson and charm quark have the same sign from the corresponding opposite sign distribution eliminates the major backgrounds (W produced in association with $b\bar{b}$ or $c\bar{c}$ pairs), which have equal probabilities to produce same- and opposite sign signatures. The measured cross sections can serve as direct input to PDF fits [26], where they help to

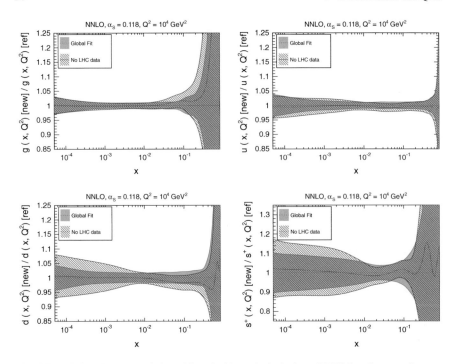

Fig. 4.7 Relative PDF uncertainties with and without the inclusion of LHC data for several proton constituents (Adapted from Ref. [26].)

constrain the strange quark density of the proton. The ratio of the cross sections with W^+ and W^- bosons provides additional information about the strange- anti-strange asymmetry in the proton.

So far, two PDF fits have included a large section of LHC results: the fit by the MMHT group [25] and by the NNPDF collaboration [26] in addition the the various electroweak measurements described above, the collaborations also include the $t\bar{t}$ production cross sections as well as jet data. The inclusion of these data considerably reduces the uncertainty on the PDFs. Figure 4.7 shows the uncertainties of the PDFs in- and excluding the LHC data, showing a considerable gain in precision. Nevertheless, substantial uncertainties remain when considering the expectations for very high and low partonic center of mass energies.

4.4   Outlook

Some LHC measurements have already been included in the latest PDF determinations [25, 26], and it is to be expected that the number of results included will rise substantially in the near future. Data from the LHC are of particular interest at large momentum fractions, when heavy quarks, or quark-antiquark differences are involved, as these were not easily accessible at previous facilities. The clear experimental signature and well understood theory of electroweak bosons, make them into an ideal probe in this field. If the progress in parton density determinations and our understanding of the electroweak boson production mechanism permits, the W boson mass may be determined in the near future with improved precision.

References

1. The ALEPH, DELPHI, L3, OPAL, SLD Collaborations, the LEP Electroweak Working Group, the SLD Electroweak and Heavy Flavour Groups, Precision electroweak measurements on the Z resonance. Phys. Rept. **427**, 257 (2006). doi:10.1016/j.physrep.2005.12.006. arXiv:hep-ex/0509008
2. The ALEPH, DELPHI, L3, OPAL Collaborations, the LEP Electroweak Working Group, Electroweak measurements in electron-positron collisions at W-boson-pair energies at LEP. Phys. Rept. **532**, 119 (2013). doi:10.1016/j.physrep.2013.07.004. arXiv:1302.3415
3. K.A. Olive et al., (Particle Data Group), Review of particle physics. Chin. Phys. C **38**, 090001 (2014). doi:10.1088/1674-1137/38/9/090001
4. R. Hamberg, W. van Neerven, T. Matsuura, A complete calculation of the order α_s^2 correction to the Drell-Yan K factor. Nucl. Phys. B **359**, 343–405 (1991). doi:10.1016/0550-3213(91)90064-5
5. C. Anastasiou et al., High precision QCD at hadron colliders: electroweak gauge boson rapidity distributions at NNLO. Phys. Rev. D **69**, 094008 (2004). doi:10.1103/PhysRevD.69.094008. arXiv:hep-ph/0312266
6. K. Melnikov, F. Petriello, Electroweak gauge boson production at hadron colliders through $\mathcal{O}(\alpha_s^2)$. Phys. Rev. D **74**, 114017 (2006). doi:10.1103/PhysRevD.74.114017. arXiv:hep-ph/0609070
7. S. Catani et al., Vector boson production at hadron colliders: a fully exclusive QCD calculation at NNLO. Phys. Rev. Lett. **103**, 082001 (2009). doi:10.1103/PhysRevLett.103.082001. arXiv:0903.2120
8. U. Baur et al., Electroweak radiative corrections to neutral current Drell-Yan processes at hadron colliders. Phys. Rev. D **65**, 033007 (2002). doi:10.1103/PhysRevD.65.033007. arXiv:hep-ph/0108274
9. U. Baur, D. Wackeroth, Electroweak radiative corrections to $p\bar{p} \rightarrow W^{\pm} \rightarrow \ell^{\pm}\nu$ beyond the pole approximation. Phys. Rev. D **70**, 073015 (2004). doi:10.1103/PhysRevD.70.073015. arXiv:hep-ph/0405191
10. G. Watt, Parton distribution function dependence of benchmark Standard Model total cross sections at the 7 TeV LHC. JHEP **1109**, 069 (2011). doi:10.1007/JHEP09(2011)069. arXiv:1106.5788
11. CMS Collaboration, Measurement of the inclusive W and Z production cross sections in pp collisions at $\sqrt{s}$ = 7 TeV. JHEP **1110**, 132 (2011). doi:10.1007/JHEP10(2011)132. arXiv:1107.4789
12. CMS Collaboration, Measurement of inclusive W and Z boson production cross sections in pp collisions at $\sqrt{s}$ = 8 TeV. Phys. Rev. Lett. **112**, 191802 (2014). doi:10.1103/PhysRevLett.112.191802. arXiv:1402.0923
13. ATLAS Collaboration, Measurement of the inclusive $W^{\pm}$ and Z/γ^* cross sections in the electron and muon decay channels in pp collisions at $\sqrt{s}$ = 7 TeV with the ATLAS detector. Phys. Rev. D **85**, 072004 (2012). doi:10.1103/PhysRevD.85.072004. arXiv:1109.5141
14. CMS Collaboration, Measurement of the differential and double-differential Drell-Yan cross sections in proton-proton collisions at $\sqrt{s}$ = 7 TeV. JHEP **1312**, 030 (2013). doi:10.1007/JHEP12(2013)030. arXiv:1310.7291

15. CMS Collaboration, Measurements of differential and double-differential Drell-Yan cross sections in proton-proton collisions at 8 TeV. Eur. Phys. J. C **75**(4), 147 (2015). doi:10.1140/epjc/s10052-015-3364-2. arXiv:1412.1115

16. CMS Collaboration, Measurement of the muon charge asymmetry in inclusive $pp \to W + X$ production at $\sqrt{s} = 7$ TeV and an improved determination of light parton distribution functions. Phys. Rev. D **90**(3), 032004 (2014). doi:10.1103/PhysRevD.90.032004. arXiv:1312.6283

17. CMS Collaboration, Measurement of the electron charge asymmetry in inclusive W production in pp collisions at $\sqrt{s} = 7$ TeV. Phys. Rev. Lett. **109**, 111806 (2012). doi:10.1103/PhysRevLett. 109.111806. arXiv:1206.2598

18. ATLAS Collaboration, Measurement of the muon charge asymmetry from W bosons produced in pp collisions at $\sqrt{s} = 7$ TeV with the ATLAS detector. Phys. Lett. B **701**, 31–49 (2011). doi:10.1016/j.physletb.2011.05.024. arXiv:1103.2929

19. CMS Collaboration, Measurement of the rapidity and transverse momentum distributions of Z bosons in pp collisions at $\sqrt{s} = 7$ TeV. Phys. Rev. D **85**, 032002 (2012). doi:10.1103/PhysRevD.85.032002. arXiv:1110.4973

20. ATLAS Collaboration, Measurement of the Z/γ^* boson transverse momentum distribution in pp collisions at $\sqrt{s} = 7$ TeV with the ATLAS detector. JHEP **1409**, 145 (2014). doi:10.1007/JHEP09(2014)145. arXiv:1406.3660

21. CMS Collaboration, Measurement of the Z boson differential cross section in transverse momentum and rapidity in protonproton collisions at 8 TeV. Phys. Lett. B **749**, 187–209 (2015). doi:10.1016/j.physletb.2015.07.065. arXiv:1504.03511

22. D0 Collaboration, Precise study of the Z/γ^* boson transverse momentum distribution in $p\bar{p}$ collisions using a novel technique. Phys. Rev. Lett. **106**, 122001 (2011). doi:10.1103/PhysRevLett.106.122001. arXiv:1010.0262

23. ATLAS Collaboration, Measurement of angular correlations in Drell-Yan lepton pairs to probe Z/γ^* boson transverse momentum at $\sqrt{s} = 7$ TeV with the ATLAS detector. Phys. Lett. B **720**, 32–51 (2013). doi:10.1016/j.physletb.2013.01.054. arXiv:1211.6899

24. S. Catani, G. Ferrera, M. Grazzini, W boson production at hadron colliders: the lepton charge asymmetry in NNLO QCD. JHEP **1005**, 006 (2010). doi:10.1007/JHEP05(2010)006. arXiv:1002.3115

25. L. Harland-Lang et al., Parton distributions in the LHC era: MMHT 2014 PDFs. Eur. Phys. J. C **75**(5), 204 (2015). doi:10.1140/epjc/s10052-015-3397-6. arXiv:1412.3989

26. NNPDF Collaboration, Parton distributions for the LHC Run II. JHEP **1504**, 040 (2015). doi:10.1007/JHEP04(2015)040. arXiv:1410.8849

27. CMS Collaboration, Measurement of associated W + charm production in pp collisions at $\sqrt{s} = 7$ TeV. JHEP **1402**, 013 (2014). doi:10.1007/JHEP02(2014)013. arXiv:1310.1138

28. CMS Collaboration, Measurement of the production cross sections for a Z boson and one or more b jets in pp collisions at $\sqrt{s} = 7$ TeV. JHEP **1406**, 120 (2014). doi:10.1007/JHEP06(2014)120. arXiv:1402.1521

29. ATLAS Collaboration, Measurement of the production of a W boson in association with a charm quark in pp collisions at $\sqrt{s} = 7$ TeV with the ATLAS detector. JHEP **1405**, 068 (2014). doi:10.1007/JHEP05(2014)068. arXiv:1402.6263

30. ATLAS Collaboration, Measurement of differential production cross-sections for a Z boson in association with b-jets in 7 TeV proton-proton collisions with the ATLAS detector. JHEP **1410**, 141 (2014). doi:10.1007/JHEP10(2014)141. arXiv:1407.3643

31. F.A. Berends et al., Multi-jet production in W, Z events at $p\bar{p}$ colliders. Phys. Lett. B **224**, 237 (1989). doi:10.1016/0370-2693(89)91081-2

32. F.A. Berends et al., On the production of a W and jets at hadron colliders. Nucl. Phys. B **357**, 32–64 (1991). doi:10.1016/0550-3213(91)90458-A

33. H. Ita et al., Precise predictions for Z + 4 jets at hadron colliders. Phys. Rev. D **85**, 031501 (2012). doi:10.1103/PhysRevD.85.031501. arXiv:1108.2229

34. C. Berger et al., Precise predictions for W + 4 jet production at the large hadron collider. Phys. Rev. Lett. **106**, 092001 (2011). doi:10.1103/PhysRevLett.106.092001. arXiv:1009.2338

35. J. Alwall et al., The automated computation of tree-level and next-to-leading order differential cross sections, and their matching to parton shower simulations. JHEP **1407**, 079 (2014). doi:10.1007/JHEP07(2014)079. arXiv:1405.0301
36. F. Caravaglios et al., A new approach to multijet calculations in hadron collisions. Nucl. Phys. B **539**, 215–232 (1999). doi:10.1016/S0550-3213(98)00739-1. arXiv:hep-ph/9807570
37. S. Catani et al., QCD matrix elements + parton showers. JHEP **0111**, 063 (2001). doi:10.1088/1126-6708/2001/11/063. arXiv:hep-ph/0109231
38. Z. Bern et al., Next-to-leading order W + 5-jet production at the LHC. Phys. Rev. D **88**(1), 014025 (2013). doi:10.1103/PhysRevD.88.014025. arXiv:1304.1253
39. S. Frixione, B.R. Webber, Matching NLO QCD computations and parton shower simulations. JHEP **0206**, 029 (2002). doi:10.1088/1126-6708/2002/06/029. arXiv:hep-ph/0204244
40. ATLAS Collaboration, Measurement of the production cross section of jets in association with a Z boson in pp collisions at $\sqrt{s} = 7$ TeV with the ATLAS detector. JHEP **1307**, 032 (2013). doi:10.1007/JHEP07(2013)032. arXiv:1304.7098
41. ATLAS Collaboration, Measurements of the W production cross sections in association with jets with the ATLAS detector, Eur. Phys. J. C **75**(2), 82 (2015). doi:10.1140/epjc/s10052-015-3262-7. arXiv:1409.8639
42. CMS Collaboration, Measurements of jet multiplicity and differential production cross sections of Z+ jets events in proton-proton collisions at $\sqrt{s} = 7$ TeV. Phys. Rev. D **91**(5), 052008 (2015). doi:10.1103/PhysRevD.91.052008. arXiv:1408.3104
43. CMS Collaboration, Differential cross section measurements for the production of a W boson in association with jets in proton-proton collisions at $\sqrt{s} = 7$ TeV. Phys. Lett. B **741**, 12–37 (2015). doi:10.1016/j.physletb.2014.12.003. arXiv:1406.7533
44. ATLAS Collaboration, Measurement of the top quark pair production cross-section with ATLAS in the single lepton channel. Phys. Lett. B **711**, 244–263 (2012). doi:10.1016/j.physletb.2012.03.083. arXiv:1201.1889
45. G. Bohm, G. Zech, *Introduction to Statistics and Measurement Analysis for Physicists* (Verl. Dt. Elektronen-Synchrotron, Hamburg, Germany, 2010)
46. G. D'Agostini, A multidimensional unfolding method based on Bayes' theorem. Nucl. Instrum. Methods A **362**, 487–498 (1995). doi:10.1016/0168-9002(95)00274-X
47. A. Höcker, V. Kartvelishvili, SVD approach to data unfolding. Nucl. Instrum. Methods A **372**, 469–481 (1996). doi:10.1016/0168-9002(95)01478-0. arXiv:hep-ph/9509307
48. CMS Collaboration, Jet production rates in association with W and Z bosons in pp collisions at $\sqrt{s} = 7$ TeV. JHEP **1201**, 010 (2012). doi:10.1007/JHEP01(2012)010. arXiv:1110.3226
49. CMS Collaboration, Measurement of the cross section and angular correlations for associated production of a Z boson with b hadrons in pp collisions at $\sqrt{s} = 7$ TeV. JHEP **1312**, 039 (2013). doi:10.1007/JHEP12(2013)039. arXiv:1310.1349

Chapter 5
Electroweak Parameters

Precision measurement of the electroweak parameters at hadron colliders suffer from substantial uncertainty due to the composite initial state compared to studies at lepton colliders [1, 2]. Nevertheless, the high luminosities and energies available at the LHC make some precision measurements viable:

- Parameters that depend on the scale of the interaction, like the effective weak mixing angle may be measured at scales not accessible at other facilities.
- The mass at the W boson is accessible at the LHC with methods similar to the ones used at the Tevatron [3].

Measurements of these parameters are of limited interest when viewed in isolation. However, due to its rigid symmetry structure, the SM predicts a large number of observable quantities governed by only a small number of underlying constants. A combined analysis of different observables connected to the same constants of nature may thus be used to check the consistency of the SM, or may hint at deviations from it.

5.1 Effective Weak Mixing Angle

The forward-backward asymmetry A_{FB} has been extensively measured at the LEP experiments [1], where it was defined as the difference of cross sections with the outgoing lepton in the opposite hemisphere as the incoming lepton and the cross sections with incoming and outgoing leptons in the same hemisphere, normalized to the sum of the cross sections. A_{FB} is directly related to the Weinberg angle θ_W, $\sin \theta_W$ determines the relative contributions of vector- and axial-vector couplings of the Z boson to the fermions. The relation becomes more complicated when higher orders are considered, an effect that is usually absorbed into the definition of the Weinberg angle which become the effective weak mixing angle $\sin^2 \theta_{\mathrm{eff}}^{\ell}$. The direct correspondence between $\sin^2 \theta_{\mathrm{eff}}^{\ell}$ and A_{FB} is only valid for on-shell Z boson production, so that the scaling dependence of the electroweak coupling structure is best studied directly in measurements of A_{FB}.

© Springer International Publishing Switzerland 2016
M.U. Mozer, *Electroweak Physics at the LHC*, Springer Tracts in Modern Physics 267, DOI 10.1007/978-3-319-30381-9_5

At first sight, it may appear that the forward backward asymmetry, A_{FB}, of the Drell-Yan process can not be measured at the LHC owing to the symmetric initial state. However, on average, quarks carry a larger momentum fraction of the proton than anti-quarks due to the presence of the valence quarks. Correspondingly, the virtual Z boson is more often produced traveling in the direction of the initial state quark. Thus A_{FB}, and indirectly $\sin^2 \theta_{eff}^\ell$, can be measured at the LHC by observing the Drell-Yan decay asymmetry as function of the direction of travel of the di-lepton system. Compared to experiments with a known initial state (e.g. LEP, SLC), the signature is diluted, as the correspondence between the initial state and the outgoing di-lepton direction is not perfect. Correcting for this dilution requires good knowledge of the proton PDFs and is a major source of uncertainty of this measurement at the LHC.

Two studies are performed in connection to A_{FB}: a measurement of $\sin^2 \theta_{eff}^\ell$ at the Z peak [5, 7] as well as a measurement of A_{FB} as function of the di-lepton invariant mass [4, 5]. In CMS $\sin^2 \theta_{eff}^\ell$ is extracted from events with a reconstructed leptonic Z decay (see Sect. 3.1) in a three-dimensional likelihood fit to the dilepton invariant mass, rapidity and decay angle. The probability density function used in the fit takes into account the detector acceptance and efficiency as well as the dilution in the asymmetry due to the limited knowledge of the initial state, allowing for a direct extraction of $\sin^2 \theta_{eff}^\ell$. In the ATLAS collaboration two dimensional templates in the dilepton mass and A_{FB} are used to similar effect. The results, $\sin^2 \theta_{eff}^\ell = 0.2287 \pm 0.0020(\text{stat}) \pm 0.0025(\text{syst})$ for CMS and $\sin^2 \theta_{eff}^\ell = 0.2308 \pm 0.0005(\text{stat}) \pm 0.0006(\text{syst}) \pm 0.0009(\text{PDF})$ for ATLAS, are compatible with previous measurements at lepton colliders [1] (see Fig. 5.1), but somewhat less precise. Already with an integrated luminosity of 1 fb^{-1}, systematic

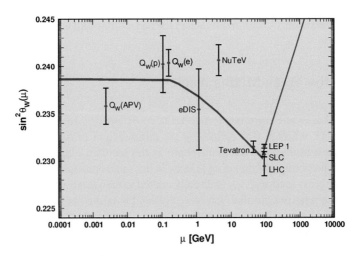

Fig. 5.1 Measurements of $\sin^2 \theta_{eff}^\ell$ as function of the hard scale. The point marked "LHC" corresponds to the combination of Refs. [4, 5]. The point labeled "Tevatron" is also measured at the Z boson mass and is shifted to the left for better visibility (Adapted from Ref. [6].)

uncertainties are larger than statistical ones. Limited knowledge of the proton PDFs translates into uncertainties in the correlation between the initial state quark direction in the di-lepton rapidity. This measurement will only indirectly profit from the very high integrated luminosities expected in the LHC Run II, in that the additional data may improve our understanding of proton PDFs and the DY production mechanism, a situation that this study shares with the measurement of the W boson mass discussed below.

The direct correspondence between A_{FB} and $\sin^2\theta_{\text{eff}}^\ell$ only exists on the Z resonance. Nevertheless, it is of interest to observe the evolution of the electroweak couplings with the interaction scale. A direct measurement of A_{FB} as function of the di-lepton invariant mass serves to demonstrate this point. The analysis [4, 5] select events with same flavor, opposite charge lepton pairs, but instead of exclusively selecting pairs close to the Z mass, events are sorted into bins of di-lepton mass. In each bin, A_{FB} is computed in the Collins-Soper frame [8], assuming that the sign of the lepton pair rapidity determines the initial state quark direction. The detector acceptance and efficiency are corrected as well as the effects of final state radiation. The results are compatible with the predictions of the SM and clearly show the evolution of A_{FB} with the interaction scale up to 1000 GeV. The additional data expected for Run II will push the range of this measurement to even higher scales.

5.2 W Boson Mass

The mass of the W boson is linked to the masses of other SM particles through loop correction. The relationships are strictly defined in the SM and its extensions, so that the simultaneous analysis of the W boson mass along with other heavy particles can be used to test the SM as discussed below. Corresponding to its importance, great effort has been expended to measure the W boson mass. The first high precision measurements were obtained at LEP by scanning the beam energy over the W boson pair production threshold and by means of full kinematic reconstruction [2]. Since then, these measurements have been surpassed in precision by measurements at the Tevatron [3] using techniques also applicable at the LHC and described in more detail below.

The hadronic W decays are not suitable for this measurement, even though the final state can be fully reconstructed, because the jet energy resolution is not sufficient to obtain competitive results. In the leptonic decay, on the other hand, the charged lepton kinematics can be reconstructed with excellent precision, while the neutrino only leaves a signature in $E_{\text{T}}^{\text{miss}}$. Even though much information is lost due to the incomplete reconstruction of the neutrino, the lepton p_{T}, $E_{\text{T}}^{\text{miss}}$ and M_{T} spectra observed in leptonic W decays depend strongly on M_{W}. The Tevatron analysis and current efforts at the LHC use this by generating templates of these spectra under different M_{W} hypothesis and fitting them to the observed spectra.

In addition to the W boson mass, the spectra also depend on the detector calibration and on the W boson decay and production processes. The proper calibration of the LHC detectors, especially for the measurement of E_T^{miss}, is laborious and slow. However, extrapolation from current results [9–11] suggests that the accuracy needed for a competitive M_W measurement can be achieved with the CMS and ATLAS detectors. In addition to the experimental uncertainties, theory uncertainties enter the measurement, mainly through the W boson production mechanism. The rapidity distribution of the W boson is related to the proton PDF. The W boson p_T distribution is governed by initial state gluon radiation. Both of these aspects of the production can only partially be accessed in perturbation theory and as a result cannot be computed from first principles. Additional complementary measurements have to be performed in order to constrain these effects, adding indirect experimental uncertainties.

The initial p_T distribution is of special concern, because it directly effects the lepton p_T and E_T^{miss}. To constrain the W p_T distribution, the p_T distribution in inclusive Z boson production is measured, as the Z boson can be well reconstructed and the two processes share similar kinematic properties. However, the W boson p_T distribution cannot directly be inferred from the Z boson distribution: Z production results from initial states with same flavor quark-antiquark pairs. Z boson production with a $b\bar{b}$ initial state, which accounts for $\sim 4\,\%$ of the total cross section, has on average several GeV higher p_T than Z bosons arising from light quark annihilation [12]. W bosons, on the other hand, are produced from different flavor pairs, with the top-bottom combination being kinematically suppressed. Due to the different initial state parton configurations, the boson p_T spectra are affected by our knowledge of the second and third generation quark densities of the proton. This issue is much less pronounced in measurements of the Tevatron, where both Z and W boson production is dominated by the valence quarks.

In order to produce a competitive W boson mass measurement with the LHC detectors it will thus be necessary to measure W and Z boson differential cross sections to improve PDF fits, measure the associated production of W and Z bosons with heavy quarks to constrain the heavy quark densities and measure precisely the Z boson p_T distribution. As discussed in Chap. 4, such measurements are being produced, but they have not yet been translated into a W boson mass measurement.

5.3 Global Fits

The rigid symmetry structure of the SM generates a large suite of observables from a small set of parameters. Independent measurements can thus be used to overconstrain the SM parameters and test it for consistency or possible deviations from the expectations. Precision measurements of Z pole observables at the ee colliders LEP and SLC together with W pair measurements at LEP build the core of the measurements used in these studies [1, 2]. They are complemented by measurements of the W boson and top quark mass obtained from the Tevatron and increasingly the LHC.

Fig. 5.2 Higher order corrections to the W^+ boson propagator induced by fermion and boson loops. Only the dominant fermion loop is shown

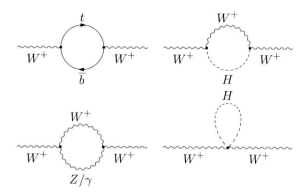

Historically, global fits to these observables have been used to predict the Higgs boson mass. This is possible because the masses of the W boson, top quark and Higgs boson are inter-related through loop corrections in their propagators. The corrections corresponding to the diagrams shown in Fig. 5.2 introduce shifts in the W boson mass that are quadratic in the top quark mass and logarithmic in the Higgs boson mass. This made it possible to reasonably constrain the mass of the top quark before its discovery, but only allowed for relatively weak constrains on the Higgs mass. The relation is demonstrated in Fig. 5.3 (left) that shows the predicted W boson mass as function of the top quark mass for different values of the Higgs boson mass.

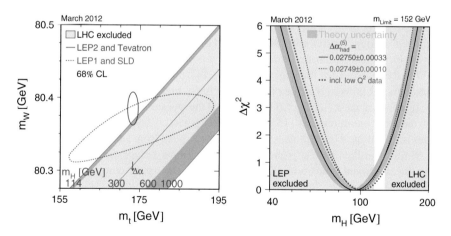

Fig. 5.3 Fits to electroweak precision measurements. *Left* 68 % confidence level contours for direct measurements of M_W and M_t from the Tevatron compared to the results for a global fit using data of LEP and SLD, as well as low energy experiments. For comparison, the SM prediction of M_W as function of M_t for a range of different Higgs boson masses are also shown. *Right* χ^2 minimum as function of the Higgs boson mass for a fit to the complete set of electroweak data discussed in the text. The *yellow regions* in the background indicate values excluded by direct search (Adapted from Ref. [1].)

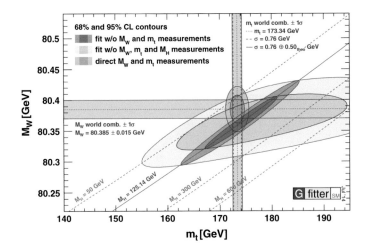

Fig. 5.4 Fit to electroweak precision measurements including the Higgs boson mass (Adapted from Ref. [15].)

The measurements of the W and top mass (shown as 68 % confidence level ellipse) can be compared to the results from the electroweak fit with the masses removed, showing overall good agreement. When experimental measurements for both, the W boson and top mass are considered, the Higgs boson mass can be constrained, but a wide range of masses remains compatible with the data, as can be seen in Fig. 5.3 (right), which shows the χ^2 as function of the Higgs boson mass of a global fit to the electroweak data, shortly before the discover of the Higgs boson. This approach predicted a Higgs boson with $M_H < 152$ GeV at 95 % confidence level [2], not considering direct exclusion limits.

With the discovery of the Higgs boson [13, 14], the global analysis of electroweak measurements turns into a true consistency check. Including recent LHC results for the Higgs boson mass (see Sect. 6.6) shows that the measured Higgs mass is consistent with the other measurements [15]. Overall, the largest contribution to the χ^2 in the fit arises from a b-qark related measurement at the Z-peak from LEP. Figure 5.4, shows the fit result in the M_t-M_W plane, demonstrating how stringently the direct Higgs mass measurement constrains the allowed parameter space compared to indirect measurements only.

The good agreement with the SM suggests that such studies can be used to constrain extensions of the SM. The most common scenario studied in such global analysis beyond the SM are various supersymmetric models (SUSY). The additional particles in these models participate in loop corrections similar to the ones shown in Fig. 5.2 and may lead to deviations in the interrelation of the particle masses compared to the SM. This behavior is, in fact, one of the attractive features of SUSY models: In the SM, the constants of nature need to be finely tuned to obtain a Higgs boson mass close to the electroweak scale, as unchecked loop corrections would naturally suggest much higher masses (the so called fine-tuning or naturalness problem).

In SUSY models on the other hand, the corrections largely cancel between SM and new particles, leading to a Higgs boson mass close to the electroweak scale. To first order, the Higgs mass is predicted to be strictly smaller than the mass of the Z boson, but higher order corrections allow Higgs boson masses up to ∼135 GeV. A global fit in terms of SUSY models allows for a direct extraction of the model parameters. This sensitivity of the Higgs boson mass to additional particles means that the measurement of the Higgs boson mass (see Sect. 6.6) provides an especially strong constraint on the possible parameter space. To improve the power of such an analysis, it is useful to extend the data set by measurements directly relevant to the new physics model under investigation. As an example, Refs. [16–18], add additional observations from flavor physics, cosmology and direct SUSY searches at the LHC as well as direct dark matter searches to gain insight into the regions of SUSY parameter space allowed by the combination of these measurements. While the allowed parameter space is greatly constrained by the measurements, a large region is still available as the number of parameters even in constrained SUSY models is significantly larger than the rather small set of SM parameters.

5.4 Outlook

With ongoing data-taking at the LHC, precision measurements of electroweak parameters are unlikely to gain substantially in precision, as the additional data will have only limited impact on the dominating systematic uncertainties. However, more precise measurements of the W boson, top quark and Higgs mass are likely to be achieved, using the improved understanding of non-perturbative effects achieved with the LHC itself (see Chap. 4). These quantities represent valuable inputs in global analyses of the SM and increased precision in their measurements may exclude (or point towards) possible SUSY scenarios in the absence of direct exclusion (or evidence).

References

1. The ALEPH, DELPHI, L3, OPAL, SLD Collaborations, the LEP Electroweak Working Group, the SLD Electroweak and Heavy Flavour Groups, Precision electroweak measurements on the Z resonance. Phys. Rept. **427**, 257 (2006). doi:10.1016/j.physrep.2005.12.006. arXiv:hep-ex/0509008
2. The ALEPH, DELPHI, L3, OPAL Collaborations, the LEP Electroweak Working Group, Electroweak measurements in electron-positron collisions at W-boson-pair energies at LEP. Phys. Rept. **532**, 119 (2013). doi:10.1016/j.physrep.2013.07.004. arXiv:1302.3415
3. CDF and D0 Collaboration, 2012 update of the combination of CDF and D0 results for the mass of the W boson, arXiv:1204.0042
4. CMS Collaboration, Forward-backward asymmetry of Drell-Yan lepton pairs in pp collisions at $\sqrt{s}$ = 7 TeV. Phys. Lett. B **718**, 752–772 (2013). doi:10.1016/j.physletb.2012.10.082. arXiv:1207.3973

5. ATLAS Collaboration, Measurement of the forward-backward asymmetry of electron and muon pair-production in pp collisions at $\sqrt{s}$ = 7 TeV with the ATLAS detector. JHEP **09**, 049 (2015). doi:10.1007/JHEP09(2015)049. arXiv:1503.03709
6. K.A. Olive et al., (Particle Data Group), Review of particle physics. Chin. Phys. C **38**, 090001 (2014). doi:10.1088/1674-1137/38/9/090001
7. CMS Collaboration, Measurement of the weak mixing angle with the Drell-Yan process in proton-proton collisions at the LHC. Phys. Rev. D **84**, 112002 (2011). doi:10.1103/PhysRevD. 84.112002. arXiv:1110.2682
8. J.C. Collins, D.E. Soper, Angular distribution of dileptons in high-energy hadron collisions. Phys. Rev. D **16**, 2219 (1977). doi:10.1103/PhysRevD.16.2219
9. CMS Collaboration, Performance of missing transverse momentum reconstruction algorithms in proton-proton collisions at $\sqrt{s}$ = 8 TeV with the CMS detector. CMS Physics Analysis Summary CMS-PAS-JME-12-002 (2013)
10. CMS Collaboration, Performance of CMS muon reconstruction in pp collision events at $\sqrt{s}$ = 7 TeV. JINST **7**, P10002 (2012). doi:10.1088/1748-0221/7/10/P10002. arXiv:1206.4071
11. CMS Collaboration, Performance of electron reconstruction and selection with the CMS detector in proton-proton collisions at $\sqrt{s}$ = 8 TeV. JINST **10**(06), P06005 (2015). doi:10.1088/1748-0221/10/06/P06005. arXiv:1502.02701
12. S. Berge, P.M. Nadolsky, F.I. Olness, Heavy-flavor effects in soft gluon resummation for electroweak boson production at hadron colliders. Phys. Rev. D **73**, 013002 (2006). doi:10.1103/PhysRevD.73.013002. arXiv:hep-ph/0509023
13. CMS Collaboration, Observation of a new boson at a mass of 125 GeV with the CMS experiment at the LHC. Phys. Lett. B **716**, 30–61 (2012). doi:10.1016/j.physletb.2012.08.021. arXiv:1207.7235
14. ATLAS Collaboration, Observation of a new particle in the search for the Standard Model Higgs boson with the ATLAS detector at the LHC. Phys.Lett. B **716**, 1–29 (2012). doi:10.1016/j.physletb.2012.08.020. arXiv:1207.7214
15. Gfitter Collaboration, The global electroweak fit at NNLO and prospects for the LHC and ILC. Eur. Phys. J. C **74**, 3046 (2014). doi:10.1140/epjc/s10052-014-3046-5. arXiv:1407.3792
16. O. Buchmueller et al., The CMSSM and NUHM1 after LHC Run 1. Eur. Phys. J. C **74**(6), 2922 (2014). doi:10.1140/epjc/s10052-014-2922-3. arXiv:1312.5250
17. C. Strege et al., Global Fits of the cMSSM and NUHM including the LHC Higgs discovery and new XENON100 constraints. JCAP **1304**, 013 (2013). doi:10.1088/1475-7516/2013/04/013. arXiv:1212.2636
18. P. Bechtle et al., Constrained supersymmetry after two years of LHC data: a global view with Fittino. JHEP **1206**, 098 (2012). doi:10.1007/JHEP06(2012)098. arXiv:1204.4199

Chapter 6
EWK Bosons and the Higgs Boson

As discussed, in Chap. 2, the Higgs boson plays an important role in the process of electroweak symmetry breaking. Correspondingly, the couplings of the electroweak boson to the Higgs boson are noticeably different in structure and strength compared to the Yukawa couplings responsible for the fermion masses. For this reason, the Higgs boson has large branching ratios to pairs of W and Z bosons, where kinematically allowed. The branching ratios are suppressed, but still substantial below the threshold for the production of two real bosons, but falls steeply for lower Higgs boson masses (see Fig. 6.1).

The preferential couplings of the Higgs boson to the W and Z boson also affect the production mechanisms at the LHC. In the SM, the dominant production channel for the Higgs boson at the LHC is the gluon fusion process (see Fig. 6.2), which can be attributed to the large gluon density in the proton and the large Yukawa couplings of the top quark. Additionally, there are significant contributions from the vector boson fusion (VBF) process. The cross section for this process falls slowly with rising Higgs bosons mass compared to the gluon fusion process, so that the relative contribution of VBF rises significantly until it reaches about 50 % for a Higgs boson mass of 1 TeV (see Fig. 6.1). A small number of Higgs bosons are also expected to be produced in association with top quarks or electroweak bosons.

Commensurate to the importance of the Higgs boson as the last undiscovered particle of the SM, the LHC experiments mounted a comprehensive set of searches for the Higgs boson in every conceivable channel. To maximize the potential for a discovery, the searches were not limited to the most promising channels, but any channel with useful sensitivity was considered, so as to gain the best results in the combination of all analysis. Above the threshold for the decay into two real W or Z bosons this includes the Higgs decay into W and Z pairs and a number of possible boson pair decay configurations. Below this threshold, the decay into $\tau^+\tau^-$, $b\bar{b}$ and $\gamma\gamma$ pairs were additionally investigated. Three of these decay channels were most instrunemtal in the discovery of the Higgs Boson: H $\to \gamma\gamma$, H $\to$ WW $\to 2\ell2\nu$ and H $\to$ ZZ $\to 4\ell$.

© Springer International Publishing Switzerland 2016
M.U. Mozer, *Electroweak Physics at the LHC*, Springer Tracts
in Modern Physics 267, DOI 10.1007/978-3-319-30381-9_6

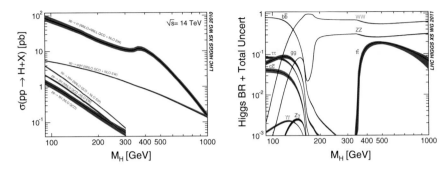

Fig. 6.1 Higgs production cross section for several production mechanisms [1] (*left*) and Higgs Branching Ratios [2] (*right*) as function of Higgs boson mass (Adapted from Refs. [1, 2].)

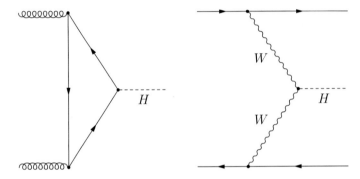

Fig. 6.2 Higgs production mechanisms. *Left* gluon fusion; *right* vector boson fusion

6.1 Higgs Searches in Di-photon Decays

As discussed in Chap. 2, the photon does not couple directly to the Higgs boson. The decay is mediated through a loop diagram similar to the gluon fusion production process. The branching fraction in this channel is correspondingly small (see Fig. 6.1). However, below the Higgs mass threshold to produce two real W or Z boson, the di-photon decay is one of the more powerful channels, due to its clear detector signature compared to the more common decays to $b\bar{b}$ or $\tau^+\tau^-$ pairs.

The Higgs searches in the di-photon final state proceed very similarly in the CMS and ATLAS experiments [3–6]: The basic strategy of the analysis is fairly simple: events with two photons are selected and the invariant mass-spectrum of the photon pairs is searched for a narrow peak on a smoothly falling background. While simple in its basic idea, the analyses are extremely complex in practice.

In addition to the irreducible background induced by SM di-photon production, this Higgs boson search suffers from background induced by π^0 decays. The signatures of the two photons of a π^0 decay will overlap in the electromagnetic calorimeter for pions with momenta comparable to Higgs-decay photons and can easily be mistaken for single photons. Ultimately, this background is much reduced by a number of selection criteria that test the shape of the electromagnetic shower of each photon candidate for compatibility with an origin from a single hard photon or a pion decay.

The di-photon final state also suffers from a unique issue not encountered in the other Higgs boson decays: due to the all neutral final state, the photons cannot easily be associated to a collision vertex. This leads to difficulties for track-based photon isolation variables, but also affects the resolution of the reconstructed invariant mass of the photon pair. The experiments overcome this problem by using multivariate methods to choose the most likely vertex based on variables such as the p_T of the photon pair compared to the p_T of the tracks emerging from a given vertex. The ATLAS experiment also uses the longitudinal segmentation in its electromagnetic calorimeter to extrapolate the photon direction, which is not possible in the homogeneous CMS ECAL.

To further optimize the senstivity of the search, the di-photon datasets are split into a number of categories along two sets of critera: On the one hand the samples are split according to additional particles in the event to separate different production mechanims. On the other hand, there is a separation along the lines of photon reconstruction quality to obtain sub-samples of high and low purity.

The search for the Higgs boson signal is performed as a simultaneous fit of parametric background shapes and a narrow signal peak in the di-photon invariant mass spectra of the various categories. Due to the small statistical uncertainties and large backgrounds in this channel, even small deviations in the background from the idealized shapes could be misinterpreted as spurious signals. To avoid such biases, both experiments perform extensive studies using a variety of possible background shapes. To ensure that the spectrum can be reasonably described by a simple parametric approach, the photon p_T threshold is set dynamically dependent on the di-photon mass, avoiding hard to describe turn-on effects at low di-photon masses. Figure 6.3, shows the resulting di-photon mass spectrum including the background and signal fits, clearly showing the excess at 125 GeV.

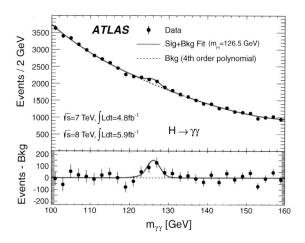

Fig. 6.3 Di-photon mass spectrum obtained by the ATLAS experiment including signal and background fits at the time of the Higgs discovery (Adapted from Ref. [4].)

6.2 Higgs Searches in WW Decays

At the beginning of the LHC data-taking, the mass of the Higgs boson, and in fact even its existence was not known. Considering the privileged interactions between the Higgs boson and the electroweak boson, special emphasis was put by the experiments on searches in the H → ZZ and H → WW channels. The analysis is complicated by the large number of different possible decay channels, particularly for the Z case, which multiplies the number of effective final states to be studied (see Fig. 6.4). At the LHC, all different final states other than 2q2ν, 4q and 4ν have been studied.

The study of Higgs decays in the diboson channel naturally segregates into two different regimes, which are treated separately in most analysis: for $M_H = 2 \cdot M_W$, the Higgs boson decays almost exclusively to two W bosons. This mass, where a H → ZZ analysis has no sensitivity marks the boundary between low mass searches (with at least one off-shell boson) and high mass searches (with two on-shell bosons).

This distinction is the least pronounced in the WW → $2\ell 2\nu$ channel [7, 8], where the two-neutrino final state hinders the reconstruction of the W masses as well as the Higgs mass. The analysis proceeds by selecting events with two oppositely charged leptons and E_T^{miss}. A major difficulty are the accurate estimate of the background rates, as the Higgs signal does not produce a narrow peak due to the two final state neutrinos. The analysis separates the data into a set of categories based on the number of jets and jet kinematics, which allows simultaneously to better control the background and study the Higgs production mechanism. In same flavor decays, a substantial amount of Drell-Yan background remains, even after excluding events with a di-lepton mass close to the nominal Z mass. The signal component is enriched by the inclusion of the opening angle between the two leptons $\Delta\phi_{\ell\ell}$ in the analysis. In signal events, $\Delta\phi_{\ell\ell}$ is small as the spin-0 Higgs boson decays into polarized W bosons and the parity violating decay of the W bosons produces aligned leptons (see Fig. 6.5). To optimize the analysis, selection criteria in the CMS analysis are not implemented as simple cuts, but instead used in the form of a multivariate discriminant. Even though the

Fig. 6.4 Cross sections times branching ratios for the H → VV processes for different final states. Only final states studied by the LHC experiments are shown (Adapted from Ref. [2].)

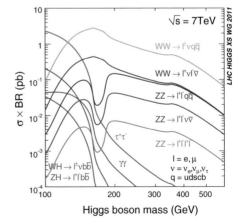

Fig. 6.5 Distribution of the opening angle between the two leptons $\Delta\phi_{\ell\ell}$ in the $H \to WW \to 2\ell2\nu$ analysis in the ATLAS experiment (Adapted from Ref. [9].)

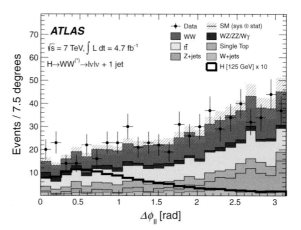

mass resolution of the channel is poor, the selection is optimized separately for a large number of Higgs mass hypothesis in order to maximize the sensitivity. The ATLAS analysis pursues a different strategy: the event selection criteria are uniform, greatly simplifying the evaluation of systematic uncertainties and the signal is searched in a likelihood fit to the kinematic distributions.

In the $WW \to \ell\nu2q$ channel, the Higgs mass can be reconstructed by reconstructing the neutrino kinematics from a W mass constraint [10, 11]. The analysis are conceptually similar to the $ZZ \to 2\ell2q$ analysis described in detail below. The major difference is the reconstruction of the leptonically decaying boson, and the fact that top quark induced backgrounds play a larger role than in the the corresponding ZZ analysis.

6.3 Higgs Searches in ZZ Decays

Superficially the $ZZ \to 2\ell2\nu$ final state may appear very similar to the $WW \to 2\ell2\nu$ final state, but significant differences appear in the kinematic properties of the final state leptons. The $ZZ \to 2\ell2\nu$ channel is studied in events with a leptonic Z candidate and E_T^{miss}. The channel is particularly sensitive for high Higgs boson masses, as the E_T^{miss} corresponds to the p_T of the outgoing Z boson, which is on average directly related to the Higgs mass as $E_T^{miss} = p_{T,Z} \sim (M_H - 2 \cdot M_Z)/2$. Conversely, a search in this channel is extremely difficult in the low mass regime, where the Z bosons are expected to carry negligible p_T so that no E_T^{miss} arises from the invisible Z decay. This is rather different to the $WW \to 2\ell2\nu$ final state, where the alignment of the W decay axes suppresses the E_T^{miss} expected for high Higgs boson masses. The analysis have poor mass resolution compared to the other ZZ final states [12, 13]. The backgrounds drop very sharply with increasing E_T^{miss}, until mostly the irreducible SM production of Z pairs remains for $E_T^{miss} \gtrsim 100$ GeV. This means, combined with the larger branching ratio compared to the 4ℓ channel, that this final state is the most sensitive for very high Higgs mass hypothesis.

The Higgs searches in the $ZZ \rightarrow 4\ell$ channel [14, 15] profit from extremely low backgrounds: SM processes with four leptons in the final state have very low cross sections and the chance to obtain leptons from misdentifications drops exponentially with the number of required leptons. However, this channel suffers from an extremely low branching ratio (0.5 % if electrons and muons are considered, 1 % if also τ leptons are included). The CMS analysis includes the τ decay channels to maximize signal acceptance, but due to the complications associated to the number different τ decay channels and escaping neutrinos, the τ channels contribute overall little to the final results, even though they represent roughly half of the branching fraction.

In order to cover the whole mass range starting from the direct exclusion limit obtained at LEP [16], it is important to maintain high acceptance and efficiency even for very low p_T leptons and considerable effort was undertaken to reliably reconstruct these leptons even with transverse momenta as low as 7 GeV for electrons and 5/6 GeV muons in CMS/ATLAS, pushing the detectors to their limits. The fully reconstructed final state gives access to the full set of decay angles, allowing improved background suppression by the use of an angular likelihood discriminant, similar to the one used in the $ZZ \rightarrow 2\ell 2q$ final state discussed below. The very high purity and good resolution are also the reason why this decay channel was one of the major contributors to this Higgs boson discovery, combined with the $WW \rightarrow 2\ell 2\nu$ and $\gamma\gamma$ decays. These properties that were helpful in the discovery as well as the absence of neutrinos in this final state makes the 4ℓ channel ideally suited for measurements of the Higgs boson properties as discussed below.

While the $ZZ \rightarrow 4\ell$ decay channel promises excellent mass resolution and very small backgrounds, it suffers from a comparatively small branching ratio of only 0.5 %. As the Higgs production cross section falls steeply with the Higgs boson mass (see Fig. 6.1), the sensitivity of the 4ℓ channel becomes limited by the branching ratio at high masses. In this regime, channels with higher branching ratios become important. The $ZZ \rightarrow 4q$ decay channel has a high branching ratio of $\sim$50 %, but its backgrounds are many orders of magnutide larger than for the $ZZ \rightarrow 4\ell$ channel. A good compromise between branching ratio and backgrounds can be acchieved in the $ZZ \rightarrow 2\ell 2q$ channel, which has a branching fraction $\sim$20 times as large the 4ℓ channel. We will discuss the analysis of this channel in some detail to highlight common features among the Higgs boson searches in VV decays.

The first Higgs search in the semileptonic ZZ decays was performed on the data taken in 2011. The ATLAS experiment was able to exclude a small mass range around $M_H \simeq 350$ GeV near the point of maximal sensitivity [17]. The corresponding CMS analysis [5] could not exclude the SM Higgs boson and presents exclusion limits on exotic models with a fourth generation of heavy fermions, which enhance the gluon fusion production process. We will discuss these analysis in more detail below to highligthing several of the techniques used in the search for well understood diboson resonances. As discussed above, the regime with a $M_H < 2 \cdot M_Z$ is kinematically quite different from the mass range where a decay to two real Z bosons is allowed, and accordingly this mass range is treated separately, also in this decay channel. Due to the low number of expected signal events, gluon fusion and VBF production processes are not treated separately.

6.3.1 Search in the High Mass Regime

The reconstruction of the leptonic Z decay closely follows the steps used in the SM Z production studies discussed in Chap. 4.3. Hadronic Z boson candidates are reconstructed from jets with $p_T > 25$ GeV and $|\eta| < 2.5$ in [17] that are at least $\Delta R < 0.4$ removed from any electron. All possible combinations of the three highest p_T jets in the event are considered, leading to the possibility of more than one candidate per event. Only pairs in the mass range $75 < M_{jj} < 105$ GeV are retained for further analysis ("signal region"), while events with a jet pair in the mass range $40 < M_{jj} < 150$ GeV (excluding the signal region) are used in background estimates ("sideband region"). Events that do not contain a jet pair in either the signal or sideband region are discarded. In order to suppress backgrounds containing the decay products of top quarks, events are required to show no significant E_T^{miss}.

To better suppress the dominant background of Z bosons produced in association with jets (Z+jets), the sample is split into four categories according to the b-tags of the hadronic Z candidate and the event kinematics. The Z boson branching fraction to b quarks (~20 % of all hadronic decays) is much larger than the fraction of Z+b events when compared to production of Z bosons in association with light jets. Thus the events are sorted into two categories, where the hadronic Z candidate contain two or fewer b-tagged jets, respectively. The untagged category contains the majority of the signal, but also large backgrounds from Z+jets production, while the 2-tag category has the higher signal purity, but low signal acceptance. In addition to the Z+jet background the b-tagged category has a significant amount of $t\bar{t}$ background. With the integrated luminosity used in this analysis, the untagged category is the most sensitive, as relatively few signal events are expected. For the search of very heavy Higgs bosons ($M_H > 300$ GeV), the decay products of the two bosons are expected to be emitted relatively close to each other due to the high momenta of the Z boson. Thus, the analysis is separated into low and high mass categories, orthogonally to the b-tag requirement, by imposing the condition $\Delta\Phi < \pi/2$ on the lepton and jet pairs. The background estimate is obtained from simulation, normalized to the observed data sidebands. The observed mass distributions and background estimates are shown in Fig. 6.6 together with the expected Higgs boson signal.

The corresponding CMS analysis [5] follows in general a similar strategy, but differs in some key point. Notably, it uses a uniform selection for all Higgs mass hypothesis and has no corresponding high- and low-mass regions. The mass resolution for the reconstruction of the Higgs boson candidate is improved by a kinematic fit of the hadronic Z, as described in Sect. 3.2. The resulting adjusted four-vectors of the jets are used for the further analysis, especially the reconstruction of the Higgs candidate mass and the decay angles.

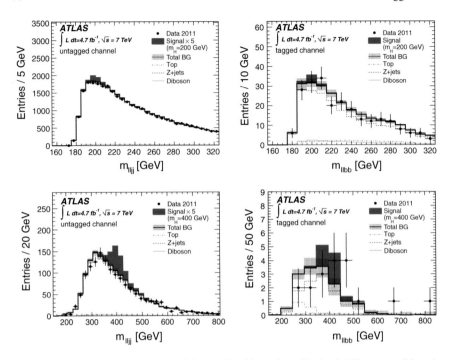

Fig. 6.6 Invariant mass distributions of the combined leptonic and hadronic Z boson candidates in two b-tag categories (*Left* 0 or 1 b-tags; *right* 2 b-tags.) and two kinematic regimes (*top* low mass; *bottom* high mass) (Adapted from Ref. [17].)

From the four-vectors of the final state particle, the complete set of five decay angles (see Fig. 6.7), can be computed with the exception of θ_2: As the quark charge cannot be determined from the jet with any certainty, only $|\cos\theta_2|$ is available. In the case of the Higgs boson, the distribution of decay angles is independent of the production mechanism and can be computed analytically [18]. This intimate knowledge of the signal distributions is exploited with a likelihood discriminant, which computes the likelihood ratio for an angular configuration to be the result of the Higgs boson decay or a background process. The background likelihoods are derived from simulation, as no analytic likelihood is available for the main background process (Z+jets). The most discriminating of the decay angles is $\cos\theta^*$ (see Fig. 6.7), which is expected to have a maximum at zero for the spin 0 signal, while showing strong bias towards small/large value for the background. While this analysis uses the decay angles to look for a signal of known quantum numbers, it is possible to turn this relation around and use them to measure the quantum numbers of an observed signal (see Sect. 6.6).

The treatment of the b-tags is also somewhat more involved, separating three categories, where the hadronic Z candidate contain two, one or zero b-tagged jets. In the 0-tag category, background is further suppressed by the removal of gluon-like jets (see Sect. 3.2).

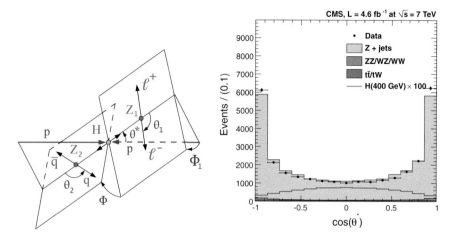

Fig. 6.7 *Left* Decay angle definition for the H → ZZ decay. *Right* Distribution of $\cos\theta^*$ in data compared the background and signal simulations (Adapted from Ref. [5].)

The background is estimated from the data in the sideband of the reconstructed mass of the hadronic Z decay before the kinematic fit. The M_{ZZ} spectrum is constructed from events in the sideband with minor backgrounds (t$\bar{\text{t}}$, SM ZZ production) subtracted. The resulting spectrum is multiplied by the ratio of the signal- and sideband-region spectra in the Z+jets simulation to correct for the subtle correlations between M_{ZZ} and M_{jj}. The corrected distribution is fitted with an empiric function, which serves as the final background estimate.

A number of systematic uncertainties affect the measurement. Uncertainties in the reconstruction of the final state particles lead to uncertainties in the expected signal yield. The expected yield is also sensitive to a number of theoretical uncertainties, such as uncertainties in the PDFs and missing higher orders in the computation. In the CMS analysis the dominant background estimate is constructed from data itself, it has only minor theory uncertainties, mainly due to the estimation of the minor backgrounds and the transfer function used in the extrapolation of the sideband. The major background uncertainties are related to the statistical uncertainties in the sideband region and corresponding uncertainty on the background fit. In the ATLAS analysis, on the other hand the leading uncertainties in the background estimate arise from uncertainties in the simulation used to model the background shapes.

The exclusion limits on the signal strength μ relative to the SM expectation are computed in the modified frequentist approach [19, 20]. Two tail probabilities are computed with the observed data: the probability to obtain a value for a test statistic q_μ larger than the observed value q_μ^{obs} for the signal+background hypothesis and for the background-only hypothesis. The ratio of these two values, CL_s gives the confidence level at which a given signal strength may be excluded, i.e. for $CL_s = 0.05$ the SM signal ($\mu = 1$) is excluded at 95 % confidence level. The test statistic q_μ is the likelihood ratio obtained from two unbinned maximum likelihood fits to the data,

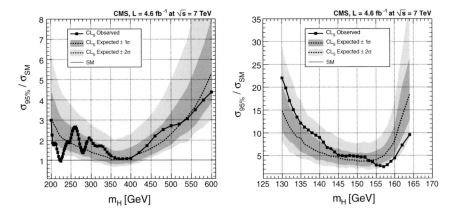

Fig. 6.8 Observed (*solid*) and expected (*dashed*) limits on the Higgs boson production cross section normalized to the SM cross section in the high mass (*left*) and low mass (*right*) regions (Adapted from Ref. [5].)

where the numerator includes a fixed signal contribution of strength μ and while the denominator is evaluated with the best fit value fo μ. Systematic uncertainties are taken into account by profiling, i.e. the are included in the likelihood fits to evaluete q_μ as free parameters, constrained by additional suitable probability densities. Figure 6.8 (left) shows the limit as observed by the CMS collaboration.

6.3.2   Search in the Low Mass Regime

The search for a Higgs boson in the $2\ell2q$ channel in the low mass regime is more difficult and less sensitive than in the high mass regime. The most common type of sharing of the Higgs bosons rest energy between the two outgoing Z bosons is to have one on-shell Z boson and one far off shell Z boson. The cross section for the final state where both Z bosons are moderately off-shell and the corresponding interference terms are comparatively small [2]. This separation into real and virtual Z bosons effectively doubles the possible final states, corresponding to the assignment of the real and virtual Z to the hadronic or leptonic decay.

This decay channel was only investigated by the CMS collaboration and in the analysis presented here, only the case of the leptonically decaying boson representing the virtual Z is pursued. The main Z+jets background rises steeply if the selection on the dijet mass is loosened to allow lower dijet masses due to the increased number of possible dijet combinations, especially if the jet p_T requirements are relaxed at the same time. The opposite is true when shifting the invariant mass selection of the lepton pair to lower masses, as the Drell Yan cross section considerably falls off the Z peak. Accordingly, the analysis for the low mass region studies di-lepton events with invariant mass <80 GeV, relaxing the p_T requirements to 20 (10) GeV for the leading (subleading) lepton in order to maintain good signal acceptance lower Higgs hypothesis masses. Nevertheless, the signal acceptance falls dramatically towards low masses, limiting the analysis of the low mass region to a range between $130 < M_H < 170$ GeV.

Other than the lepton selection, the low mass analysis follows the example of the high mass analysis described above with the exception of the angular discriminant, which show little separation power in this regime and is omitted. Figure 6.8 (right) shows the corresponding exclusion limits, which rise steeply at low M_H due to the limited acceptance. This limited acceptance also prevents follow-up studies on later discovered H(125) in this channel: The increased instantaneous luminosity in the LHC running period of 2012 demands higher trigger thresholds for the lepton triggers, so this type of analysis has very little signal acceptance.

6.4 Current Status of Higgs Searches

In the Summer of 2012, significant amounts of excess events beyond background expectations were observed in the more sensitive channels [4, 21]. At that time, no single channel could produce a clear discovery by itself, but the particle physics community was quickly convinced of the presence of a new particle compatible with the Higgs boson by means of the good consistency of many aspects of observed excesses, which were consistent between different search channels, corroborated by both experiments and matching SM Higgs boson expectations, including the results of electroweak precision fits discussed in Sect. 5.3. A more detailed account of the discovery may be found in Ref. [22]. As the new particle is consistent in its properties with the SM Higgs boson it will be referred to as "the" Higgs boson or H(125), referring to its mass in GeV.

Since the discovery of the Higgs boson in 2012, the integrated luminosity taken with the CMS and ATLAS detectors has more than doubled. This additional data has been used to perform a number of studies on the Higgs boson beyond its mere discovery. The three channels that contributed most significantly to the discovery ($H \rightarrow \gamma\gamma$, $H \rightarrow ZZ \rightarrow 4\ell$ and $H \rightarrow WW \rightarrow 2\ell2\nu$) have been used to measure the properties of the new particle and its compatibility with the SM Higgs boson as described in more detail in Sect. 6.6. The intervening time has also given opportunity to the experiments to combine their results, increasing the effective luminosity and mutually constraining systematic uncertainties, as can be seen in the combined measurement of the Higgs boson mass [23]. The discovery of the Higgs boson also signalled the beginning of a large effort to search for an extended Higgs sector and similar exotic phenomena involving the Higgs boson, as described in Sects. 6.5 and 7.4.

The additional luminosity has also enabled the experiments to find evidence for the Higgs boson in additional decay channels. Foremost among these is the decay to $\tau^+\tau^-$ pairs [24, 25]. These measurements are the first direct demonstration that the new particle couples to leptons and that the couplings are consistent with a SM Higgs boson. This final state also plays a major role in the measurements of the Higgs couplings discussed below, as the $\tau^+\tau^-$ channel is the only directly observed decay to fermions. Beyond its relevance in the SM, the $\tau^+\tau^-$ results are also interpreted in SUSY models, where the decay to $\tau^+\tau^-$ pairs plays in important role.

The decays to the lighter leptons have been studied [26, 27], but much more luminosity is needed before the decay of a SM Higgs boson to muons may be observed. The muon channel provides the most realistic opportunity to verify the Higgs boson couplings to second generation fermions, as an analysis with strange- or charm quarks in the final state are expected to be even more challenging than the study of decays to bottom quarks discussed below. As the branching ratio of the Higgs boson to fermions scales with the square of the corresponding Yukawa coupling, and thus with the square of the fermion mass, the branching ratio to muons is suppressed by a factor $\sim$300 compared to the decay to $\tau^+\tau^-$ pairs. However, the sensitivity is reduced by a much smaller factor as the reconstruction efficiency and precision is larger for the muon channel, while backgrounds are lower. An observation of the decay to electrons is unfeasible for the SM Higgs as the branching ratio is expected to be a mere 5×10^{-9}. Even over the whole lifetime of the LHC the SM signal in this channel cannot be expected to be observed and and the channel is studied in the terms of exotic models with modified couplings instead.

The coupling of the Higgs boson to quarks is also an active area of study. The decay of the Higgs boson to $b\bar{b}$ pairs is the dominant decay mode for a SM Higgs boson of mass 125 GeV, but the analysis is hampered by the very large background of $b\bar{b}$ pair production through QCD processes. To reduce this background and increase sensitivity, this decay channel is studied in Higgs bosons produced in association with vector bosons. Due to the low cross section of this process, the $b\bar{b}$ decay has not yet been firmly established [28, 29]. The studies are particularly sensitive in the kinematic regime, where the Higgs boson and recoiling vector boson have high transverse momentum. To identify these energetic H → $b\bar{b}$ decays, techniques similar to the reconstruction of boosted hadronic vector boson decays (see Sect. 3.2) are employed, adding b-tag criteria to the discriminating variables.

The coupling between the Higgs boson and top quark is of interest due to the prominent role of the top quark corrections to the Higgs boson mass. First constraints were derived from measurements of the Higgs gluon-fusion production cross section and di-photon decays, which receive loop-contributions involving the top quark. Nevertheless, due to the model-dependence of these estimates, a direct measurement is desirable. While the decay of the Higgs boson to top quarks is kinematically forbidden, the coupling can be directly studied in the associated production of a Higgs boson with top quarks. This process has a low cross section as well, so that at the moment only limits on its magnitude have been published [31–33]. While such measurements are sensitive to the absolute value of the top quark Yukawa coupling, it would be necessary to measure the production of Higgs bosons in association with a single top quark to directly measure the sign of the coupling. However, due to interference effects, the cross section for this process is extremely low and measurements sensitive to the SM cross sections are not expected in the near future.

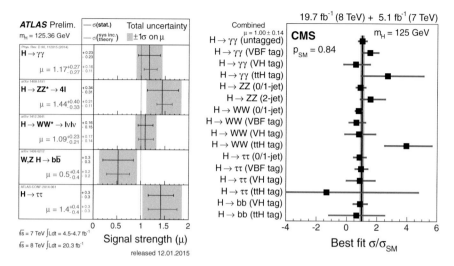

Fig. 6.9 Summary of Higgs boson measurements by the ATLAS (*left*, Source: ATLAS/CERN) and CMS (*right*, Adapted from Ref. [30]) experiments

The measurements of the various Higgs boson decay channels are summarised in Fig. 6.9. Beyond the fundamental measurements of the existence and strength of the expected Higgs boson decays, the additional luminosity has also enabled the extraction of differential cross sections in [34, 35], giving insight into the details of the production mechanism.

6.5 Search for Exotic Scalar Bosons

With the discovery of the Higgs boson at a mass of 125 GeV [4, 21], the various Higgs searches were re-evaluated in the light of this discovery. Broadly two directions were taken: channels with good sensitivity to the H(125) boson or channels of specific interest in Higgs phyiscs (for example teh decay to $\tau^+\tau^-$ pairs discussed above) are studied to measure the properties of the new particle. Others, especially the searches in the high mass regime were re-cast in terms or searches for extensions of the SM with a extended Higgs sector. These include a wide range of WW and ZZ final states [36], but also di-photon pairs [37]. With a Higgs boson candidate identified, decays of new particles into Higgs bosons now present well defined signatures and a number of searches for such particles have been undertaken and will be discussed in Sect. 7.4.

We will continue to follow the $ZZ \rightarrow 2\ell2q$ channel as an example for a search of additional scalar bosons at high mass. Studies of the H(125) in the $ZZ \rightarrow 2\ell2q$ channel were found to be inferior to equivalent studies using the $ZZ \rightarrow 4\ell$ channel. The search for further scalar resonances, on the other hand, proved much more promising: with an extension of the mass range of the previous analysis, the large branching fraction of the ZZ system into the 2l2q final state and steeply falling backgrounds allow for competitive results. Such additional resonance could appear if there is an additional electroweak singlet [38, 39] as it could arise in models where dark matter is connected to observable matter through the Higgs sector [38]. Measurements using this interpretation complement and compete with direct dark matter detection experiments. Alternatively additional Higgs bosons also appear in models with a second Higgs doublet, as is common in SUSY models [40, 41]. Similar searches for additional scalar bosons are also performed in other bosonic decay channels and combined in Ref. [36].

The search of a high mass scalar particle in the data taken in 2012 [42] follows broadly the analysis strategy of the the previous analysis [5]. Nevertheless, several conceptual improvements were added:

- The analysis is sensitive to the production mechanism by tagging forward jets indicative of VBF production.
- In order be sensitive to boson up to a mass of 1 TeV, boosted Z decays are considered.

In addition, several smaller improvements are included in this analysis, for example the estimate of the $t\bar{t}$ background from a control sample with oppositely charged $e\mu$ pairs and a more sophisticated treatment of pile-up. With the increased integrated luminosity of the 2012 data taking period (roughly four times the 2011 data), the most sensitive category changes to the sample with two b-tags, due to it's good purity.

6.5.1 Boosted Topologies

The initial Higgs search in the 2l2q channel [5] covered Higgs masses up to 600 GeV. With the focus on the search for additional exotic resonances and the higher integrated luminosity of the 2012 data taking period, an extended mass range was seen as attractive. As discussed in Sect. 3.2, the two jets from the hadronic Z decay are expected to overlap in this extended mass range.

Accordingly, the analysis of the 2012 data set splits the data sample into two subsamples: A "boosted" and a "dijet" category. All events are subject to a common selection criteria concerning the leptonic Z decay. The selection criteria are very similar to the analysis on the 2011 data set, but lepton p_T thresholds are increased to 40 (20) GeV for the leading (subleading) lepton and isolation variables are corrected for pile-up contributions. If a given event has a leptonic Z with transverse momentum above 200 GeV and a jet with more than 100 GeV, it is considered as part of the boosted category and otherwise in the dijet category, assuming that the Higgs decay products will be roughly balanced in p_T. The exact selection procedure and category precedence has very little effect on the final results, as both categories have similar sensitivity in the kinematic region where an event may qualify for either category. For events in the dijet category, analysis proceeds as described above. The highest

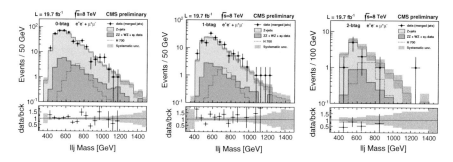

Fig. 6.10 Invariant mass spectra o the two lepton + jet system in events with a leptonic Z and V-tagged jet. *Left* no subjet btag; *middle* one subjet b-tag; *right* two subjet b-tags (Adapted from Ref. [42].)

p_T jet in a boosted event is required to have a pruned mass close to the Z boson mass and a low N-subjettiness in order to enrich the sample in boosted Z decays. Additionally, sub-jet b-tagging is applied to maximize the sensitivity of the analysis (see Fig. 6.10).

Compared to the di-jet category, systematic uncertainties are somewhat increased for the boosted decay channel. The simulated τ_{21} spectrum differs significantly from the one observed in data-control samples (see Sect. 3.2). The discrepancy is corrected by a scale factor, but an uncertainty of 10 % is estimated to remain. Notably increased is also the uncertainty associated to the sub-jet b-tag efficiency and mis-identification rate, as described in Sect. 3.2. Even though systematic uncertainties are somewhat increased, the inclusion of the boosted Z decays allows for an inrease in the reach of the analysis up to boson masses of 1 TeV.

6.5.2 VBF Tagging

The additional outgoing quarks of the VBF process (see Fig. 6.1) can be used to separate this topology from other Higgs boson production mechanisms. The additional jets formed from these quarks are expected to be widely separated in η and have a large invariant mass between them. Specifically investigating the VBF process by studying the additional jets originating from the outgoing quarks of the VBF process is a common feature among the Higgs searches discussed here and brings two advantages:

- The presence of the additional jets can be used to suppress backgrounds, so that the sensitivity of the analysis can be increased, especially at high mass, where the VBF contribution to the signal is large.
- The ratio of production cross sections in gluon fusion and VBF is a prediction of the SM and is potentially modified in models beyond the SM. In the case of a discovery, the signal yields observed for the different production mechanisms becomes a discriminating variable to test the SM against more exotic theories.

In addition to the hadronic Z candidate, which is composed of jets within $|\eta| < 2.4$, jets with $|\eta| < 4.7$ are considered for the tagging of events produced in the VBF process. The absence of track information used in the PF algorithm in this detector region degrades the performance of jet energy calibration and pile-up corrections. Nevertheless, the vastly increased acceptance for the tagging jets motivates the inclusion of these very forward jets. If an event contains two such jets in addition to the hadronic Z candidate, and the tagging pair also fulfills the condition that the separation of the tagging jets exceeds $\Delta\eta > 3.5$ and their invariant mass $M_{jj} > 500$ GeV, the event is considered to be produced in the VBF process.

Selection criteria on the reconstructed Higgs boson candidate closely follow the selection described above, with the exception that the cut on the angular likelihood discriminant is relaxed. This looser requirement improves the sensitivity in this event category, which has a lower expected number of signal events but also higher purity than the untagged categories described above.

Additionally, a multivariate discriminant is employed to measure the compatibility of the tagging jets with the VBF process and reject events from the production of Z bosons in association with three or four jets. The discriminant uses the following variables:

- the separation $\Delta\eta$ and $\Delta\phi$ between the tagging jets, invariant mass of the tagging jet pair;
- tagging jet energies, transverse momenta and pseudorapidities;

Events are required to pass a threshold, corresponding to a signal efficiency of 70–80 % (20 %) for events produced in the VBF (gluon fusion) process, respectively. The background is suppressed by $\sim$90 % at this working-point.

Passing events are analyzed in two categories, the electron and muon channels. Due to the low number of events in this channel, subdivisions into more channels (i.e. with b-tags, merged vs. dijet Z reconstruction) would lead to large statistical fluctuations in various data and simulated control samples, increasing uncertainties.

The VBF category suffers from additional uncertainties related for the tagging jets: the efficiency to reconstruct VBF events depends on the calibration of the forward calorimeters with corresponding uncertainties. Additionally, there are migration uncertainties related to the cross section of Higgs production in association with two jets in a gluon fusion process. This signal contribution may fall into the acceptance of the VBF analysis, but has to be considered as gluon induced when studying the Higgs couplings. The corresponding cross sections and kinematics are computed at NLO using the MINLO technique [43], but nevertheless the uncertainties are substantial, partially because for most of the mass range studied, the gluon fusion cross section is much larger than the VBF cross section. The corresponding uncertainties due to migration of VBF events into the non-tagged categories is much lower, as the VBF cross section and kinematics are well known and VBF cross section is low.

When these results are combined with the other diboson channels discussed above, an additional boson with SM Higgs boson like properties can be excluded up to masses of 1 TeV (see Fig. 6.11). With the addition of the boosted Z decay category in particular, the sensitivity of the $2\ell 2q$ final state surpasses the 4ℓ final state for

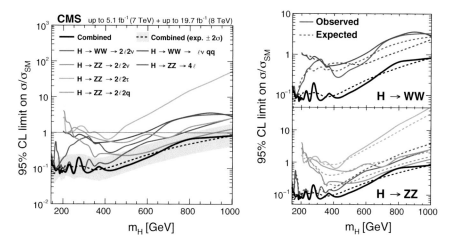

Fig. 6.11 Combined limit on a Higgs boson like particle at high mass, using WW and ZZ decays (Adapted from Ref. [36].)

boson mass hypothesis larger than ∼700 GeV. The exclusion limits on an additional electroweak singlet are more stringent than limits obtained from measurements of the H(125) peak for masses up to ∼700 GeV, depending on the model assumptions [36].

6.6 Higgs Boson Properties

Several properties of the H(125) boson may be measured by studying its decay to vector bosons. Especially the H → ZZ → 4ℓ decay is used due to its exceptional purity:

- At the most basic level, the production cross section multiplied by the branching fraction to two electroweak bosons is directly available from the count of observed signal candidates [7, 8, 14, 34, 44].
- In combination with other decay channels and the tagging of production mechanisms, the event yields may be used to extract relative couplings of the H(125) to fermions and bosons [30, 45].
- The position of the peak in the invariant mass spectrum of the 4ℓ system indicates the mass of the Higgs boson [23, 30, 46].
- While a very loose limit on the Higgs width may be obtained from the direct observation of the peak in the 4ℓ spectrum, much more stringent limits can be found from measurements of the the interference of the tail of the Higgs boson mass distribution with high mass SM ZZ production [47, 48].
- The forward backward asymmetry of the vector boson decays can be used to infer the spin states of the decay bosons and allows for corresponding test of possible quantum numbers of the new boson [49–51].

Fig. 6.12 Measured
couplings of the Higgs boson
as function of particle mass,
note the different scales for
fermions (Yukawa
couplings) and vector bosons
(square-root of the coupling
for the HVV vertex divided
by twice the vacuum
expectation value of the
Higgs field) (Adapted from
Ref. [30].)

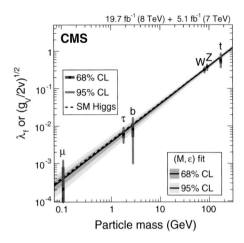

The measurement of cross sections and branching ratios are especially interesting when they are used to extract the couplings to vector bosons as well as fermions, as this allows to verify the role of the H(125) in the generation of particle masses. As shown in Fig. 6.12, measured coupling strengths are compatible with being proportional to the particle masses, as described in Sect. 2.1. While the results are compatible with the SM, the interpretation of deviations remains ambiguous, as the extraction is firmly based on SM assumptions, so that larger deviations would not be straight forward to interpret [30, 45].

The mass of the H(125) boson can be well measured in the 4ℓ channel, as the decay is fully reconstructed with excellent resolution. Only the H $\rightarrow \gamma\gamma$ decay channel shows a similarly good mass resolution. As an additional advantage the, 4ℓ invariant mass spectrum shows a prominent peak at the Z mass from radiative Z boson decays, which serves as a fixed point to calibrate the mass scale. Even with the full Run I dataset, the mass measurement in the 4ℓ channel is still limited by statistical uncertainties and prospects for Run II point to notably improved accuracy in the mass measurement. The currently available measurement already has a major impact on the viability of supersymmetric scenarios [52] and improved mass determinations will further constrain this sector.

In the SM a Higgs boson with mass $\sim$125 GeV is expected to have a width a a few MeV, orders of magnitude lower than the width that could be measured by a direct fit to the peak shape in the invariant mass distribution, which is of the order of a few GeV. However, assuming the that H(125) is indeed the SM Higgs boson, the relative cross section ratio for on-shell and far off-shell higgs boson production is sensitive to the width [53, 54]. Such a measurement is possible in the H $\rightarrow$ ZZ $\rightarrow 4\ell$ and $2\ell2\nu$, as well as H $\rightarrow$ WW $\rightarrow 2\ell2\nu$ channels, because the off-shell cross-section is resonantly enhanced above the two-boson threshold, where interference with the Higgsless SM gg $\rightarrow$ VV processes produces measurable deviations in the diboson production cross section. With this method, the Higgs boson width can be constrained to within a factor $\sim$5 of the SM expectation [47, 48].

The spin correlations of the the decay bosons have been used to enrich signal processes with vacuum quantum numbers as described above. However, the principle may be reversed: if the signal sample is selected without optimizing the selection to scalar bosons, the angular distributions can instead be used to determine the quantum numbers of the observed resonance. Again the H → ZZ→ 4ℓ is particularly suitable due to the high purity of the channel and the full kinematic reconstruction of the final state. The WW → 2ℓ2ν and diphoton final states are also used, but don't carry as rich information. With the amount of data available from Run I, a complete partial wave analysis of the 4ℓ final state is not feasible. Instead likelihood discriminants are constructed to test specific alternative quantum number hypothesis against the SM [49–51]. While this analysis allows the exclusion of a number of alternatives, such a pseudoscalar resonances, the method is cumbersome in the study of spin-2 models, as the large number of possible spin states of the resonance, depending on the production mechanism, multiplies the number of hypothesis to be tested. With the luminosity expected from Run II, a more holistic approach will probably become feasible.

6.7 Outlook

Data from Run II of the LHC will be very useful in further studies of the Higgs boson. Especially the particularly pure H → ZZ → 4ℓ decay has a very low branching fraction (0.5 % of all ZZ final states) so that current measurements can still be improved with larger data sets. This should allow an even more precise determination of the Higgs boson mass. In addition, it should become possible to perform a proper partial wave analysis of the Higgs boson decay and directly measure its quantum numbers instead of the current approach of building a number of discriminating variables to exclude a given set of alternative models. In the search for additional high mass scalars, the LHC Run II promises very large improvements, as the parton luminosity at the high invariant masses under study will rise disproportionally at high masses due to the increased beam energy. Compared to the published results of Run I, however, it may be expected that future studies may be performed in a less model specific manner in order to gain sensitivity to a wider variety of alternative models.

References

1. S. Dittmaier et al., (LHC Higgs Cross Section Working Group), Handbook of LHC Higgs Cross Sections: 1. Inclusive Observables. doi:10.5170/CERN-2011-002. arXiv:1101.0593
2. S. Dittmaier et al., (LHC Higgs Cross Section Working Group), Handbook of LHC Higgs Cross Sections: 2. Differential Distributions. doi:10.5170/CERN-2012-002. arXiv:1201.3084
3. ATLAS Collaboration, Measurement of Higgs boson production in the diphoton decay channel in pp collisions at center-of-mass energies of 7 and 8 TeV with the ATLAS detector. Phys. Rev. D **90**(11), 112015 (2014). doi:10.1103/PhysRevD.90.112015. arXiv:1408.7084

4. ATLAS Collaboration, Observation of a new particle in the search for the Standard Model Higgs boson with the ATLAS detector at the LHC. Phys. Lett. B **716**, 1–29 (2012). doi:10.1016/j.physletb.2012.08.020. arXiv:1207.7214

5. CMS Collaboration, Search for a Higgs boson in the decay channel H to ZZ(*) to $q\bar{q}\ell^-\ell^+$ in pp collisions at $\sqrt{s} = 7$ TeV. JHEP **1204**, 036 (2012). doi:10.1007/JHEP04(2012)036. arXiv:1202.1416

6. CMS Collaboration, Observation of the diphoton decay of the Higgs boson and measurement of its properties. Eur. Phys. J. C **74**(10), 3076 (2014). doi:10.1140/epjc/s10052-014-3076-z. arXiv:1407.0558

7. CMS Collaboration, Measurement of Higgs boson production and properties in the WW decay channel with leptonic final states. JHEP **1401**, 096 (2014). doi:10.1007/JHEP01(2014)096. arXiv:1312.1129

8. ATLAS Collaboration, Observation and measurement of Higgs boson decays to WW* with the ATLAS detector. Phys. Rev. D **92**(1), 012006 (2015). doi:10.1103/PhysRevD.92.012006. arXiv:1412.2641

9. ATLAS Collaboration, Search for the Standard Model Higgs boson in the $H \rightarrow$ WW(*) $\rightarrow \ell\nu\ell\nu$ decay mode with 4.7 /fb of ATLAS data at $\sqrt{s} = 7$ TeV. Phys. Lett. B **716**, 62–81 (2012). doi:10.1016/j.physletb.2012.08.010. arXiv:1206.0756 and auxiliary material at http://atlas.web.cern.ch/Atlas/GROUPS/PHYSICS/PAPERS/HIGG-2012-04/

10. CMS Collaboration, Search for a standard-model-like Higgs boson with a mass in the range 145 to 1000 GeV at the LHC. Eur. Phys. J. C **73**, 2469 (2013). doi:10.1140/epjc/s10052-013-2469-8. arXiv:1304.0213

11. ATLAS Collaboration, Search for the Higgs boson in the $H \rightarrow WW \rightarrow$ lnujj decay channel at $\sqrt{s} = 7$ TeV with the ATLAS detector. Phys. Lett. B **718**, 391–410 (2012). doi:10.1016/j.physletb.2012.10.066. arXiv:1206.6074

12. CMS Collaboration, Search for the standard model Higgs boson in the H to ZZ to $2\ell2\nu$ channel in pp collisions at $\sqrt{s} = 7$ TeV. JHEP **1203**, 040 (2012). doi:10.1007/JHEP03(2012)040. arXiv:1202.3478

13. ATLAS Collaboration, Search for a Standard Model Higgs boson in the H $\rightarrow ZZ \rightarrow l^+l^-\nu\bar{\nu}$ decay channel using 4.7 fb^{-1} of $\sqrt{s} = 7$ TeV data with the ATLAS detector. Phys. Lett. B **717**, 29–48 (2012). doi:10.1016/j.physletb.2012.09.016. arXiv:1205.6744

14. CMS Collaboration, Measurement of the properties of a Higgs boson in the four-lepton final state. Phys. Rev. D **89**(9), 092007 (2014). doi:10.1103/PhysRevD.89.092007. arXiv:1312.5353

15. ATLAS Collaboration, Measurements of Higgs boson production and couplings in the four-lepton channel in pp collisions at center-of-mass energies of 7 and 8 TeV with the ATLAS detector. Phys. Rev. D **91**(1), 012006 (2015). doi:10.1103/PhysRevD.91.012006. arXiv:1408.5191

16. The ALEPH, DELPHI, L3, OPAL Collaborations, the LEP Electroweak Working Group, Electroweak measurements in electron-positron collisions at W-boson-pair energies at LEP. Phys. Rept. **532**, 119 (2013). doi:10.1016/j.physrep.2013.07.004. arXiv:1302.3415

17. ATLAS Collaboration, Search for a standard model Higgs boson in the mass range 200–600-GeV in the $H \rightarrow ZZ \rightarrow \ell^+\ell^-q\bar{q}$ decay channel with the ATLAS detector. Phys. Lett. B **717**, 70–88 (2012). doi:10.1016/j.physletb.2012.09.020. arXiv:1206.2443

18. Y. Gao et al., Spin determination of single-produced resonances at hadron colliders. Phys. Rev. D **81**, 075022 (2010). doi:10.1103/PhysRevD.81.075022. arXiv:1001.3396

19. A.L. Read, Presentation of search results: the CL(s) technique. J. Phys. G **28**, 2693–2704 (2002). doi:10.1088/0954-3899/28/10/313

20. T. Junk, Confidence level computation for combining searches with small statistics. Nucl. Instrum. Methods A **434**, 435–443 (1999). doi:10.1016/S0168-9002(99)00498-2. arXiv:hep-ex/9902006

21. CMS Collaboration, Observation of a new boson at a mass of 125 GeV with the CMS experiment at the LHC. Phys. Lett. B **716**, 30–61 (2012). doi:10.1016/j.physletb.2012.08.021. arXiv:1207.7235

22. R. Wolf, *The Higgs Boson Discovery at the Large Hadron Collider*, vol. 264 (Springer, 2015)

23. ATLAS, CMS Collaboration, Combined measurement of the Higgs boson mass in pp collisions at $\sqrt{s} = 7$ and 8 TeV with the ATLAS and CMS experiments. Phys. Rev. Lett. **114**, 191803 (2015). doi:10.1103/PhysRevLett.114.191803. arXiv:1503.07589

24. ATLAS Collaboration, Evidence for the Higgs-boson Yukawa coupling to tau leptons with the ATLAS detector. JHEP **04**, 117 (2015). doi:10.1007/JHEP04(2015)117. arXiv:1501.04943

25. CMS Collaboration, Evidence for the 125 GeV Higgs boson decaying to a pair of τ leptons. JHEP **05**, 104 (2014). doi:10.1007/JHEP05(2014)104. arXiv:1401.5041

26. CMS Collaboration, Search for a standard model-like Higgs boson in the $\mu^+\mu^-$ and e^+e^- decay channels at the LHC. Phys. Lett. B **744**, 184–207 (2015). doi:10.1016/j.physletb.2015.03.048. arXiv:1410.6679

27. ATLAS Collaboration, Search for the Standard Model Higgs boson decay to $\mu^+\mu^-$ with the ATLAS detector. Phys. Lett. B **738**, 68–86 (2014). doi:10.1016/j.physletb.2014.09.008. arXiv:1406.7663

28. ATLAS Collaboration, Search for the $b\bar{b}$ decay of the Standard Model Higgs boson in associated $(W/Z)H$ production with the ATLAS detector. JHEP **01**, 069 (2015). doi:10.1007/JHEP01(2015)069. arXiv:1409.6212

29. CMS Collaboration, Search for the standard model Higgs boson produced in association with a W or a Z boson and decaying to bottom quarks. Phys. Rev. D **89**(1), 012003 (2014). doi:10.1103/PhysRevD.89.012003. arXiv:1310.3687

30. CMS Collaboration, Precise determination of the mass of the Higgs boson and tests of compatibility of its couplings with the standard model predictions using proton collisions at 7 and 8 TeV. Eur. Phys. J. C **75**(5), 212 (2015). doi:10.1140/epjc/s10052-015-3351-7. arXiv:1412.8662

31. ATLAS Collaboration, Search for the associated production of the Higgs boson with a top quark pair in multilepton final states with the ATLAS detector. Phys. Lett. B **749**, 519–541 (2015). doi:10.1016/j.physletb.2015.07.079. arXiv:1506.05988

32. CMS Collaboration, Search for a standard model Higgs boson produced in association with a top-quark pair and decaying to bottom quarks using a matrix element method. Eur. Phys. J. C **75**(6), 251 (2015). doi:10.1140/epjc/s10052-015-3454-1. arXiv:1502.02485

33. ATLAS Collaboration, Search for $H \rightarrow \gamma\gamma$ produced in association with top quarks and constraints on the Yukawa coupling between the top quark and the Higgs boson using data taken at 7 TeV and 8 TeV with the ATLAS detector. Phys. Lett. B **740**, 222–242 (2015). doi:10.1016/j.physletb.2014.11.049. arXiv:1409.3122

34. ATLAS Collaboration, Measurements of the total and differential Higgs boson production cross sections combining the $H \rightarrow \gamma\gamma$ and $H \rightarrow ZZ^* \rightarrow 4\ell$ decay channels at $\sqrt{s} = 8$ TeV with the ATLAS detector. Phys. Rev. Lett. **115**(9), 091801 (2015). doi:10.1103/PhysRevLett.115.091801. arXiv:1504.05833

35. ATLAS Collaboration, Measurements of fiducial and differential cross sections for Higgs boson production in the diphoton decay channel at $\sqrt{s} = 8$ TeV with ATLAS. JHEP **09**, 112 (2014). doi:10.1007/JHEP09(2014)112. arXiv:1407.4222

36. CMS Collaboration, Search for a Higgs boson in the mass range from 145 to 1000 GeV decaying to a pair of W or Z bosons. JHEP **10**, 144 (2015). doi:10.1007/JHEP10(2015)144. arXiv:1504.00936

37. ATLAS Collaboration, Search for scalar diphoton resonances in the mass range 65 − 600 GeV with the ATLAS detector in pp collision data at $\sqrt{s} = 8$ TeV. Phys. Rev. Lett. **113**(17), 171801 (2014). doi:10.1103/PhysRevLett.113.171801. arXiv:1407.6583

38. B. Patt, F. Wilczek, Higgs-field portal into hidden sectors, arXiv:hep-ph/0605188

39. V. Barger et al., LHC phenomenology of an extended standard model with a real scalar singlet. Phys. Rev. D **77**, 035005 (2008). doi:10.1103/PhysRevD.77.035005. arXiv:0706.4311

40. G. Branco et al., Theory and phenomenology of two-Higgs-doublet models. Phys. Rept. **516**, 1–102 (2012). doi:10.1016/j.physrep.2012.02.002. arXiv:1106.0034

41. N. Craig, S. Thomas, Exclusive signals of an extended Higgs sector. JHEP **1211**, 083 (2012). doi:10.1007/JHEP11(2012)083. arXiv:1207.4835

42. CMS Collaboration, Search for a standard model like Higgs boson in the H to ZZ to $\ell^+\ell^- q\bar{q}$ decay channel at $\sqrt{s}=8$ TeV, CMS Physics Analysis Summary CMS-PAS-HIG-14-007 (2015)

43. J.M. Campbell et al., W and Z bosons in association with two jets using the POWHEG method. JHEP **1308**, 005 (2013). doi:10.1007/JHEP08(2013)005. arXiv:1303.5447

44. ATLAS Collaboration, Fiducial and differential cross sections of Higgs boson production measured in the four-lepton decay channel in pp collisions at $\sqrt{s}$=8 TeV with the ATLAS detector. Phys. Lett. B **738**, 234–253 (2014). doi:10.1016/j.physletb.2014.09.054. arXiv:1408.3226

45. ATLAS Collaboration, Measurements of Higgs boson production and couplings in diboson final states with the ATLAS detector at the LHC. Phys. Lett. B **726**, 88–119 (2013). doi:10.1016/j.physletb.2014.05.011, doi:10.1016/j.physletb.2013.08.010. arXiv:1307.1427

46. ATLAS Collaboration, Measurement of the Higgs boson mass from the $H \rightarrow \gamma\gamma$ and $H \rightarrow ZZ^* \rightarrow 4\ell$ channels with the ATLAS detector using 25 fb^{-1} of pp collision data. Phys. Rev. D **90**(5), 052004 (2014). doi:10.1103/PhysRevD.90.052004. arXiv:1406.3827

47. CMS Collaboration, Constraints on the Higgs boson width from off-shell production and decay to Z-boson pairs. Phys. Lett. B **736**, 64 (2014). doi:10.1016/j.physletb.2014.06.077. arXiv:1405.3455

48. ATLAS Collaboration, Constraints on the off-shell Higgs boson signal strength in the high-mass ZZ and WW final states with the ATLAS detector. Eur. Phys. J. C **75**(7), 335 (2015). doi:10.1140/epjc/s10052-015-3542-2. arXiv:1503.01060

49. CMS Collaboration, Constraints on the spin-parity and anomalous HVV couplings of the Higgs boson in proton collisions at 7 and 8 TeV. Phys. Rev. D **92**(1), 012004 (2015). doi:10.1103/PhysRevD.92.012004. arXiv:1411.3441

50. ATLAS Collaboration, Evidence for the spin-0 nature of the Higgs boson using ATLAS data. Phys. Lett. B **726**, 120–144 (2013). doi:10.1016/j.physletb.2013.08.026. arXiv:1307.1432

51. ATLAS Collaboration, Determination of spin and parity of the Higgs boson in the $WW^* \rightarrow e\nu\mu\nu$ decay channel with the ATLAS detector. Eur. Phys. J. C **75**(5), 231 (2015). doi:10.1140/epjc/s10052-015-3436-3. arXiv:1503.03643

52. O. Buchmueller et al., The CMSSM and NUHM1 after LHC Run 1. Eur. Phys. J. C **74**(6), 2922 (2014). doi:10.1140/epjc/s10052-014-2922-3. arXiv:1312.5250

53. N. Kauer, G. Passarino, Inadequacy of zero-width approximation for a light Higgs boson signal. JHEP **1208**, 116 (2012). doi:10.1007/JHEP08(2012)116. arXiv:1206.4803

54. F. Caola, K. Melnikov, Constraining the Higgs boson width with ZZ production at the LHC. Phys. Rev. D **88**, 054024 (2013). doi:10.1103/PhysRevD.88.054024. arXiv:1307.4935

Chapter 7
Diboson Resonances

Considering the large quantum corrections to the Higgs boson mass (see Sect. 5.1), the question arises whether the observed values for the mass of the Higgs boson is the result of finely tuned constants of nature or whether yet undiscovered physical principles naturally produce the observed values. The issue is related to the observation that the electroweak scale and the corresponding Higgs- and electroweak boson masses are extremely small compared to the Planck scale, where the unification of all four forces may be expected, an issue often referred to a hierarchy problem.

The difference in observed scales between electroweak processes and the Planck scale may be related to the existence of compact extra dimensions [1]. A common feature of these extra-dimension models are the existence of a so-called tower of Kaluza-Klein (KK) excitations of the mediator of gravity, the KK graviton. Depending on the details of the model, decays of the KK graviton to vector boson pairs are possible. A long standing benchmark model in this category is the Randall-Sundrum (RS) warped extra dimension model [2]. To cure problems of the RS model in the flavor sector, it may be extended into so called "bulk graviton" models [3–5], which have been a primary target of searches with the LHC experiments.

Other extensions of the SM may also predict resonant production of electroweak boson pairs. These resonances may for example take the form of additional heavy gauge bosons [6, 7]. While models that predict new bosons largely analogous to the existing W and Z bosons are most likely to be observed in leptonic decays, there are models [8] in which fermionic decays are suppressed, increasing the importance of searches in the diboson final state. While somewhat disfavored by the recent H(125) discovery, resonances decaying to dibosons also appear in technicolor models [9].

Models that predict W' and Z' bosons can be phenomenologically subsumed into the heavy vector triple (HVT) model [10]. In this model an additional triplet of spin-1 particles is included in an effective Lagrangian, to derive the cross sections as well as decay widths for the W' and Z' particles as function of just two parameters related to the couplings of the new particles to the fermion and boson sectors.

© Springer International Publishing Switzerland 2016
M.U. Mozer, *Electroweak Physics at the LHC*, Springer Tracts
in Modern Physics 267, DOI 10.1007/978-3-319-30381-9_7

The LHC experiments have searched for such resonances in a wide range of final states, testing and excluding a variety of SM extensions. The following presents an overview of diboson resonance searches, showing a detailed example in the semi-leptonic channel.

7.1   Searches with Fully Leptonic Decays

Searches in the fully leptonic final states profit from the low backgrounds in final states with many leptons. The purity is highest for the $ZZ \rightarrow 4\ell$ final state and somewhat lower for final states that include an increasing number of neutrinos, though these channels also often have higher branching fractions, somewhat compensating for the lower purity. In the $ZZ \rightarrow 4\ell$, $ZZ \rightarrow 2\ell 2\nu$ and $WW \rightarrow 2\ell 2\nu$ channels, the searches have been motivated by models of additional bosons, mixing with the observed H(125) boson [11], similar to the studies described in Sect. 6.5. Taken together, these searches provide a powerful tool to constrain possible extensions of the Higgs sector, complementary to precision measurements of the H(125). While the combined results [11] focus in the interpretation in terms of an electroweak singlet mixing with the H(125) [12, 13], future studies may be expected to study diboson resonances under the paradigm of a second Higgs doublet [14, 15]. Such doublet models are inspired by supersymmetric theories, where the additional doublet is necessary to maintain the supersymmetry in the presence of the Higgs boson.

Additionally, the LHC experiments have searched for resonances decaying into a WZ pair in fully leptonic decays [16, 17]. This final state may arise from additional heavy charged vector bosons (W'), such as grand unified theories or models with extra dimensions [6, 18]. Many of these models are most easily detected in $\ell\bar{\nu}_\ell$ decays, though some models (for example [8]) are notably fermiphobic and show an enhanced decay to the WZ final state. Alternatively, WZ resonances may appear in technicolor models of electroweak symmetry breaking [9].

The studies select events with two opposite charge, same flavor leptons consistent with Z mass and an additional lepton as well as substantial $E_{\mathrm{T}}^{\mathrm{miss}}$. The W boson mass constraint is used to deduce the neutrino longitudinal momentum, so that the diboson mass can be computed. The signal is expected to appear as a peak in the M_{WZ} spectrum above the smoothly falling background of SM WZ production. No excess above the expectations are observed and W' bosons corresponding to the extended gauge model [6] are excluded up to masses of $\sim$1.5 TeV by both experiments. Additionally, the ATLAS collaboration interprets the results in terms of the HVT model, while CMS provides additional limits on technicolor scenarios.

Graviton decays have also been searched in the WW channel with a fully leptonic decays [19]. Even though the two neutrinos in the final state lead to a poor mass resolution, the resulting limits are quite stringent, due to the low background in this channel. Backgrounds are especially low in the final state with two different flavor leptons, where the Drell-Yan process can not contribute.

7.2 Searches with Fully Hadronic Decays

The advantages and limitations of searches in the all hadronic decay channels are opposite to those in the fully leptonic channels: the all hadronic final state has a very large branching ratio, but the background from QCD multi-jet processes is enormous. Additionally, the mass resolution of di-jet systems and the groomed mass resolution for boosted decays is not sufficient to properly distinguish hadronic W and Z decays, complicating the analysis. Commonly, a single set of selection requirements is designed so that the resulting spectra can be interpreted with a Z- and W boson hypothesis. Nevertheless, competitive results can be obtained at very high invariant masses, as the spectrum of QCD multi-jet processes falls very steeply. The situation is best in the kinematic regime of boosted hadronic boson decays, as the probability to observe jets with boson-like mass and substructure is considerably lower than observing a jet pair of roughly the W- or Z boson mass.

Representative analyses of this type are presented in Refs. [20] and [21]. The studies proceed very similarly, using events with two high p_T jets, one or both of which are identified as boosted boson decays, following the methods described in Sect. 3.2: the CMS analysis uses the pruned jet mass and N-subjettiness, while the ATLAS analysis uses the mass-drop technique in combination with a momentum balance requirement. The dataset with one V-tagged jet is sensitive to models of excited quarks [22, 23], decaying into a SM quark and an electroweak boson, while the dataset with two boson tags is used to set limits on diboson resonances, such as the graviton and W′ discussed above. While the analysis have good sensitivity at high mass, they are limited by the large multi-jet background at lower masses. An additional constraint is imposed on the analysis by the experiments trigger systems: as jet-substructure variables were not used in the CMS and ATLAS trigger systems during the LHC Run I, simple jet triggers with high p_T thresholds had to be used, limiting the analysis to resonances above 1/1.3 TeV for CMS/ATLAS, respectively. The situation may improve in the LHC Run II, where jet-substructure variables will be used in the HLT. Intriguingly, the ATLAS experiment observes a notable excess of events at an invariant mass close to 2 TeV (see Fig. 7.1). Though the excess is not significant, it implies that diboson resonances should be closely watched during Run II of the LHC.

Additionally, diboson resonances may be searched in events with one jet in association with E_T^{miss}. Usually this final state is interpreted as associated production of some dark matter candidate [24, 25], but it may also arise from a resonance decaying to a Z and another vector boson with the Z decaying to neutrinos and the other boson decaying hadronically in a boosted configuration. An initial CMS analysis [26] combines the final state of a single jet and E_T^{miss} with the semi-leptonic decay channel. The analysis is limited by the use of the jet mass as sole V-tagging mechanism, leading to higher backgrounds than analysis using more sophisticated substructure techniques.

Fig. 7.1 Invariant mass spectrum of W-tagged jet pairs in the search for a diboson resonance (Adapted from Ref. [21].)

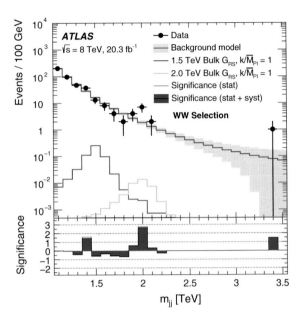

7.3 Searches with Semi-Leptonic Decays

As for the Higgs searches, the semi-leptonic decay channels offer a good compromise between background levels and branching fraction. For searches for very high resonance masses in the ZZ channel, the situation is particularly advantageous due to the low branching fraction of the Z to leptons and the quickly falling background as function diboson invariant masses. This can be clearly seen in a combined analysis of the $ZZ \rightarrow 2l2q$ and $ZZ \rightarrow 4\ell$ channels [27], where the combined limit in the search for a heavy graviton is dominated by the semileptonic channel, while the limits on gravitons with mass less than $\sim$500 GeV are driven by the 4ℓ channel. Fully leptonic decays are selected by requiring two well reconstructed leptonic Z decays, as described in Sect. 3.1. Already this basic selection produces a sample that is practically free of reducible backgrounds and SM ZZ production processes dominate the selected sample. In the semileptonic channel a single leptonic Z decay is combined with jet pairs with an invariant mass close the the Z mass. To reduce the large backgrounds from the production of single Z boson in association with jets, the analysis is optimized for two regions of resonance masses, where the higher mass region requires higher p_T of the boson candidates. Possible signals are expected to manifest as narrow peaks above the smoothly falling background in the diboson invariant mass spectrum. A similar analysis by the CMS experiment [28] focusing on the semi-leptonic channel only will be discussed in more detail below.

Resonances decaying to W-pairs in the semileptonic channel were initially studied by the ATLAS experiment [29] in conjunction with WZ resonances, as the dijet mass resolution is not sufficient to cleanly separate the two channels. As only one neutrino is involved in this final state, the full event kinematics can be reconstructed by using the W mass as a constraint as described in Sect. 3.1. The hadronic boson is constructed from jet pairs close to the W and Z mass. Backgrounds are estimated from simulation, normalized to appropriate control regions: a sideband in the dijet mass for the dominating W+jet background and a sample with b-tagged jets to determine the subleading $t\bar{t}$ background. The resulting spectra are interpreted in terms of a graviton (in the WW interpretation) and a W′ (in the WZ interpretation).

The early (i.e. 2011) analysis in the semileptonic channel are all limited in their mass reach by the jet merging phenomenon discussed in Sect. 3.2, as all of them are based on resolved boson decays only. For the 2012 data-taking period, both experiments used V-tagging techniques to extend the mass reach of their searches substantially [30, 31]. The ATLAS analysis uses the splitting scale (see Sect. 3.2) to tag boosted boson decays and combines boosted and resolved channels to set limits on graviton production in WW decays as well as W′ production in WZ decays. The corresponding CMS analysis is focused entirely on the boosted decay channels and combines the WW and ZZ decays of gravitons. In addition, it presents model independent limits, parameterized as function of resonance mass and width, in contrast to the earlier CMS analysis, which heavily relied on model-specific optimizations. In the following we will discuss the two CMS studies in more detail.

7.3.1 Graviton Search with Resolved Jets

The CMS semileptonic ZZ resonance search using data taken in 2011 [28] largely follows the corresponding Higgs analysis (see Sect. 6.3). The leptonic and hadronic Z reconstruction are independent of the intermediate resonance and are employed without change. The major complication arises from the angular likelihood discriminant. While the background likelihood remains unchanged, a new likelihood function and acceptance parameterization are necessary for the signal. The Higgs boson, as a CP-even spin-0 particle, cannot carry information about the production process, so that a single likelihood function will describe all CP-even spin-0 resonances. In contrast, the spin-2 graviton carries a complex polarization state, inducing variations in the final state angular distributions depending on the production mechanism. Thus, the use of the likelihood discriminant allows for a highly sensitive, but also highly model dependent analysis.

As the main focus of the analysis, the likelihood discriminant is reworked for the Bulk Graviton model [3]. Nevertheless, some of the angular distributions, especially $\cos \theta^*$, do not strongly depend on the details of the model, so that the efficiency for other graviton models is high as well. The well known RS Graviton [2] was chosen as additional model to provide more widely applicable results. Figure 7.2 shows the the distribution of $\cos \theta^*$, which is the most powerful single variable in the discriminant also for the graviton, as well as the resulting discriminant. A Bulk Graviton and RS Graviton hypothesis are shown in comparison to the background simulation, showing that the discriminant selects RS graviton events well, even though it is optimized for the Bulk Graviton case.

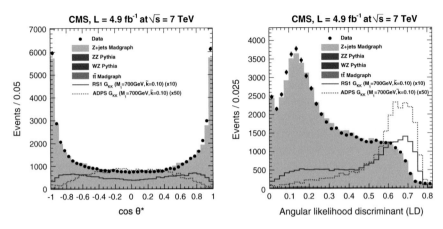

Fig. 7.2 Distribution of cos θ^* (*left*) and the angular likelihood discriminant (*right*) for the data, as well as background and signal simulations (Adapted from Ref. [28].)

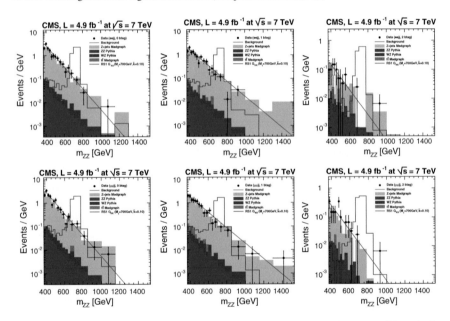

Fig. 7.3 Invariant mass spectra of the combined leptonic and hadronic Z boson candidates in the electron channel (*top*) and muon channel (*bottom*), divided into b-tag categories (*Left* 0 b-tag; *middle* 1 b-tag; *right* 2 b-tags). The background estimate from the sideband is shown as *red line*. The *filled histograms* represent the simulation and are shown for illustration (Adapted from Ref. [28].)

Selecting events with high values of the likelihood discriminant, spectra of M_{ZZ} are formed in categories split by the lepton flavor and number of b-tagged jets in the hadronic boson decay candidate (see Fig. 7.3). The background is estimated from events with the di-jet mass somewhat off the Z peak, similar to the corresponding Higgs boson analysis. The background estimated from data agrees well with the simulation and is dominated by the production of Z bosons in association with jets. No significant deviations from the expectations are observed, so limits on the RSG and Bulk Graviton production cross section are derived. At high graviton mass hypothesis, the power of the analysis is limited by the two jets of the hadronic decay merging.

7.3.2 Generic Search with Boosted Hadronic Decays

This analysis [31], using data taken during 2012 and corresponding to $19.7\,\mathrm{fb}^{-1}$, extends the search described above in three major aspects:

- The reach in resonance mass is greatly extended by focusing the analysis on boosted hadronic decays.
- WW and ZZ final states are analyzed in parallel to obtain the combined statistical power of these two channels.
- Limits are derived in a model independent manner as function of resonance mass and width. An efficiency parameterization is provided to allow the interpretation of the results in terms of a generic resonance hypothesis.

The event selection is similar to the other semi-leptonic diboson studies discussed above, with some notable exceptions. As the analysis is designed to probe very high invariant masses, up to 2.5 TeV, the effects of the boost becomes important also for the leptonic Z decays. At the highest momenta of the Z, the two decay leptons will often be so close together that they lie within each others isolation areas (see Sect. 1.2.1), leading the the rejection of these events should the common isolation variables be used. To prevent this, the isolation variables are corrected for the presence of the second lepton. This is easily possible during the offline analysis of the data, but was not foreseen in the di-lepton triggers. Non-isolated single-lepton triggers are used instead. The higher p_T thresholds of these triggers are not a limiting factor here, as the final state is highly energetic. The close spatial proximity of the outgoing leptons is also problematic for the muon reconstruction, as the CMS muons system is not optimized for the measurements of two very closeby tracks. To reduce the impact of this effect, the analysis proceeds using pairs of one global muon and one tracker muon (see Sect. 1.3.2).

These effects do not occur in the WW channel, where only one lepton is present. The leptonic W boson is reconstructed as described in Sect. 3.1, using $E_\mathrm{T}^{\mathrm{miss}}$ and the W mass constraint, to reconstruct the full kinematic quantities of the event. In this channel, the p_T threshold applied to the leptonic boson candidate is substantially higher than in the Z case to suppress backgrounds from multi-jet events with a misidentified lepton.

The reconstruction of the hadronic V decay is similar to the corresponding Higgs boson search (see Sect. 6.5): jets are reconstructed with the Cambridge-Aachen algorithm with a radius parameter $R = 0.8$ and selected based on the pruned mass, though no subjet b-tags are used. The hadronic boson candidates are required to have $p_\mathrm{T} > 200$ GeV to reduce background from SM electroweak boson production in association with jets. As the analysis reaches very high momenta of the hadronically decaying boson, the efficiency of a simple cut on τ_{21} decreases as the the subjets move closer together and are difficult to separate. To avoid this fall in efficiency, the analysis is performed in two categories of τ_{21}: the so called "high purity" category with $\tau_{21} < 0.5$ is rich in signal in all kinematic regions, while the "low purity" category contains a large amount of background and little signal at low diboson masses with a substantial signal fraction at very high diboson masses.

Figure 7.4 shows the invariant mass spectra of the diboson system for the high purity categories in data compared to simulation as well as the background estimate from the sideband in M_{pruned}. The dominant V+jets background falls exponentially, allowing for very stringent exclusion limits at high resonance masses.

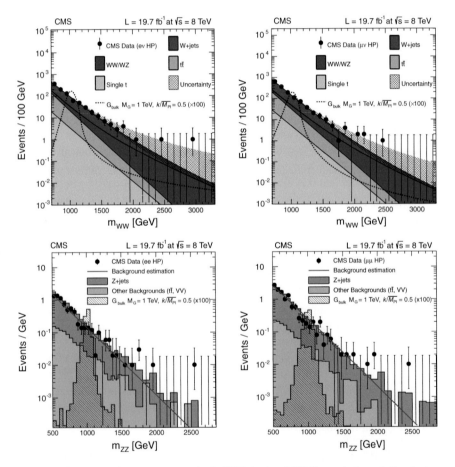

Fig. 7.4 Diboson invariant mass spectra in the WW (*top*) and ZZ (*bottom*) channel, the electron channel is shown on the *left*, the muon channel on the *right*. For the WW-channel, the W+jets background fitted to the sideband data is shown in addition to minor backgrounds estimated from simulation. For the ZZ channel, the total background estimate is shown in addition to the background simulation (Adapted from Ref. [31].)

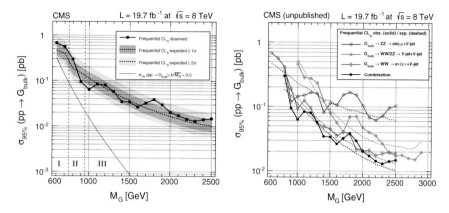

Fig. 7.5 Observed exclusion limits for a Bulk Graviton, combining the WW and ZZ decay channels in the semi-leptonic and fully hadronic final states (*left*). The *right panel* shows the different contributions from the semi-leptonic ZZ and WW channels as well has the fully hadronic channel (Adapted from Ref. [31].)

The limit on the cross section of a potential resonance is evaluated using the modified frequentist approach [32, 33], similar to the other analysis presented here. The presence of a specific model being tested allows for the combination of the WW and ZZ final states, as the relative branching fractions are known as a prediction of the model, here a Bulk Graviton model [3]. To maximize the discovery potential, the semi-leptonic results are combined with an analysis in the all-hadronic final state [20]. The results are shown in Fig. 7.5. When comparing the different channels, it can be seen that the semi-leptonic ZZ channel is most competitive for low graviton masses, simply because the high purity of the leptonic Z reconstruction allows the analysis to proceed, where the semi-leptonic WW channel is limited by trigger- and kinematic requirements that suppress the background. Over most of the mass range, the semi-leptonic WW channel is the most sensitive one, having a significantly higher branching ratio than the semi-leptonic ZZ and lower background than the fully hadronic channels.

In order to allow a broader interpretation of the results, model independent limits are also derived. The simulation of a narrow resonance is used to extract the M_{VV} resolution as function of the resonance mass. The resolution is modeled as a Gaussian core with powerlaw tails and extracted separately for the WW and ZZ final states as well as the muons and electron channels. The WW channel has somewhat worse resolution than the ZZ category due to the presence of E_T^{miss} in the reconstruction. At low invariant masses, electron and muon channel resolution are comparable, but in the TeV range, the resolution of the muon channel degrades as the track curvature decreases, while the calorimetric measurement of the electron remains precise. The signal is then modeled as the convolution of a relativistic Breit-Wigner function convoluted with the resolution and the result checked with a set of representative simulations.

Limits are then computed using these signal models as function of the resonance mass and width, spanning relative widths from 0 to 40%. The WW and ZZ channels cannot be combined in this approach as this would require knowledge of the relative branching ratio into the WW and ZZ channels. Similarly, the generic limit is restricted to the high purity category, as relative contributions to the low- and high purity categories depend slightly on the decay kinematics. In order to interpret the resulting exclusion limit, it is necessary to compute the efficiency for the generic resonance. For this purpose, the detector efficiency has been parameterized as function of the outgoing boson and lepton kinematics as well as boson polarization, so that the efficiency for any resonance can be computed as long as the decay kinematics are known. The systematic uncertainty associated to the parameterization of the efficiency is included in the generic limit. Thus the generic limit is slightly less powerful than the one tuned to the Bulk Graviton.

7.4   Searches Involving the Higgs Boson

Theories that posit new particles decaying into Higgs bosons have become more well defined with the discovery of the H(125) boson and the observation that its properties are compatible with the SM Higgs boson. Such theories include models with an additional Higgs doublet, which would manifest as a set of four additional heavy bosons: a pseudoscalar boson A with a large branching ratio for the decay $A \to HZ$, a heavy scalar boson, which may decay into a pair of Higgs bosons and two charged bosons with decays to the WH final state [34]. In models with excited vector bosons W', the decay $W' \to HW$ may occur. While the $H \to b\bar{b}$ decay has played only a minor role in the discovery of Higgs boson due to the large SM $b\bar{b}$ pair backgrounds, it is used in many resonance searches due to the high branching fraction and the large background suppression possible when studying boosted $H \to b\bar{b}$ decays with jetsubstructure techniques. Studies in this sector have not been as well organized as the searches for the SM Higgs boson discussed in Sect. 7.5 or the one for gravitons discussed above, so that the studied final states account for a somewhat random subset of the available decay channels. So far, searches in the WH and ZH channel with leptonic electroweak boson decays and $H \to b\bar{b}$ and $H \to \tau^+\tau^-$ [35–37] have been performed, as well as a search in the $HZ \to \tau^+\tau^-qq$ channel in the boosted regime [38]. The VH final state has also been studied in the fully hadronic decay, using jet substructure techniques to identify the hadronically decaying bosons [39]. The additional charged bosons expected from a second Higgs doublet have been searched in the terms of a WZ resonance, using the semi-leptonic decay channel [40]. Searches for a resonance decaying to two H(125) bosons are available in the final state with four b-jets [41, 42], two boosted Higgs jets [42] and the $HH \to \gamma\gamma b\bar{b}$ final state [43].

7.5 Outlook

Searches for new physics in multi-boson final states are just at their beginning. The power of jet substructure techniques has allowed to use high branching fraction final states for these searches, significantly increasing the reach of the LHC. These techniques are still young and future improvements may still enhance the sensitivity of such studies. With the increased center-of-mass energy and luminosity expected from the LHC Run II, many more exciting results will be forthcoming. With an improved understanding of the SM backgrounds it may also become feasible to move away from resonance searches and constrain new physics from the distortions it may introduce in the diboson mass spectrum.

References

1. N. Arkani-Hamed, S. Dimopoulos, G. Dvali, The hierarchy problem and new dimensions at a millimeter. Phys. Lett. B **429**, 263–272 (1998). doi:10.1016/S0370-2693(98)00466-3. arXiv:hep-ph/9803315
2. L. Randall, R. Sundrum, A large mass hierarchy from a small extra dimension. Phys. Rev. Lett. **83**, 3370–3373 (1999). doi:10.1103/PhysRevLett.83.3370. arXiv:hep-ph/9905221
3. K. Agashe et al., Warped gravitons at the LHC and beyond. Phys. Rev. D **76**, 036006 (2007). doi:10.1103/PhysRevD.76.036006. arXiv:hep-ph/0701186
4. A.L. Fitzpatrick et al., Searching for the Kaluza-Klein graviton in bulk RS models. JHEP **0709**, 013 (2007). doi:10.1088/1126-6708/2007/09/013. arXiv:hep-ph/0701150
5. O. Antipin, D. Atwood, A. Soni, Search for RS gravitons via W(L)W(L) decays. Phys. Lett. B **666**, 155–161 (2008). doi:10.1016/j.physletb.2008.07.009. arXiv:0711.3175
6. G. Altarelli, B. Mele, M. Ruiz-Altaba, Searching for new heavy vector bosons in $p\bar{p}$ colliders. Z. Phys. C **45**, 109 (1989). doi:10.1007/BF01552335, doi:10.1007/BF01556677
7. C. Grojean, E. Salvioni, R. Torre, A weakly constrained W' at the early LHC. JHEP **1107**, 002 (2011). doi:10.1007/JHEP07(2011)002. arXiv:1103.2761
8. H.-J. He et al., CERN LHC signatures of new gauge bosons in minimal higgsless model. Phys. Rev. D **78**, 031701 (2008). doi:10.1103/PhysRevD.78.031701. arXiv:0708.2588
9. E. Eichten, K. Lane, Low-scale technicolor at the tevatron and LHC. Phys. Lett. B **669**, 235–238 (2008). doi:10.1016/j.physletb.2008.09.047. arXiv:0706.2339
10. D. Pappadopulo et al., Heavy vector triplets: bridging theory and data. JHEP **1409**, 060 (2014). doi:10.1007/JHEP09(2014)060. arXiv:1402.4431
11. CMS Collaboration, Search for a Higgs boson in the mass range from 145 to 1000 GeV decaying to a pair of W or Z bosons. JHEP **10**, 144 (2015). doi:10.1007/JHEP10(2015)144. arXiv:1504.00936
12. B. Patt, F. Wilczek, Higgs-field portal into hidden sectors, arXiv:hep-ph/0605188
13. V. Barger et al., LHC phenomenology of an extended standard model with a real scalar singlet. Phys. Rev. D **77**, 035005 (2008). doi:10.1103/PhysRevD.77.035005. arXiv:0706.4311
14. G. Branco et al., Theory and phenomenology of two-Higgs-doublet models. Phys. Rept. **516**, 1–102 (2012). doi:10.1016/j.physrep.2012.02.002. arXiv:1106.0034
15. N. Craig, S. Thomas, Exclusive signals of an extended Higgs sector. JHEP **1211**, 083 (2012). doi:10.1007/JHEP11(2012)083. arXiv:1207.4835
16. CMS Collaboration, Search for new resonances decaying via WZ to leptons in proton-proton collisions at $\sqrt{s} = 8$ TeV. Phys. Lett. B **740**, 83–104 (2015). doi:10.1016/j.physletb.2014.11.026. arXiv:1407.3476

17. ATLAS Collaboration, Search for WZ resonances in the fully leptonic channel using pp collisions at $\sqrt{s} = 8$ TeV with the ATLAS detector. Phys. Lett. B **737**, 223–243 (2014). doi:10. 1016/j.physletb.2014.08.039. arXiv:1406.4456
18. J.L. Hewett, T.G. Rizzo, Low-energy phenomenology of superstring inspired E(6) models. Phys. Rept. **183**, 193 (1989). doi:10.1016/0370-1573(89)90071-9
19. ATLAS Collaboration, Search for new phenomena in the WW to $\ell\nu\ell'$ ν' final state in pp collisions at $\sqrt{s} = 7$ TeV with the ATLAS detector, Phys. Lett. B **718**, 860–878 (2013). doi:10.1016/j.physletb.2012.11.040. arXiv:1208.2880
20. CMS Collaboration, Search for massive resonances in dijet systems containing jets tagged as W or Z boson decays in pp collisions at $\sqrt{s} = 8$ TeV. JHEP **1408**, 173 (2014). doi:10.1007/ JHEP08(2014)173. arXiv:1405.1994
21. ATLAS Collaboration, Search for high-mass diboson resonances with boson-tagged jets in proton-proton collisions at $\sqrt{s} = 8$ TeV with the ATLAS detector. JHEP **12**, 055 (2015). doi:10.1007/JHEP12(2015)055. arXiv:1506.00962
22. U. Baur, I. Hinchliffe, D. Zeppenfeld, Excited quark production at hadron colliders. Int. J. Mod. Phys. A **2**, 1285 (1987). doi:10.1142/S0217751X87000661
23. U. Baur, M. Spira, P. Zerwas, Excited quark and lepton production at hadron colliders. Phys. Rev. D **42**, 815–824 (1990). doi:10.1103/PhysRevD.42.815
24. CMS Collaboration, Search for dark matter, extra dimensions, and unparticles in monojet events in proton-proton collisions at $\sqrt{s} = 8$ TeV. Eur. Phys. J. C **75**(5), 235 (2015). doi:10.1140/ epjc/s10052-015-3451-4. arXiv:1408.3583
25. ATLAS Collaboration, Search for new phenomena in final states with an energetic jet and large missing transverse momentum in pp collisions at $\sqrt{s} = 8$ TeV with the ATLAS detector. Eur. Phys. J. C **75**(7), 299 (2015). doi:10.1140/epjc/s10052-015-3517-3, doi:10.1140/epjc/s10052-015-3639-7. arXiv:1502.01518 [Erratum: Eur. Phys. J. C **75**(9), 408(2015)]
26. CMS Collaboration, Search for exotic resonances decaying into WZ/ZZ in pp collisions at $\sqrt{s} = 7$ TeV. JHEP **1302**, 036 (2013). doi:10.1007/JHEP02(2013)036. arXiv:1211.5779
27. ATLAS Collaboration, Search for new particles decaying to ZZ using final states with leptons and jets with the ATLAS detector in $\sqrt{s} = 7$ TeV proton-proton collisions. Phys. Lett. B **712**, 331–350 (2012). doi:10.1016/j.physletb.2012.05.020. arXiv:1203.0718
28. CMS Collaboration, Search for a narrow spin-2 resonance decaying to a pair of Z vector bosons in the semileptonic final state. Phys. Lett. B **718**, 1208–1228 (2013). doi:10.1016/j.physletb. 2012.11.063. arXiv:1209.3807
29. ATLAS Collaboration, Search for resonant diboson production in the WW/WZ→ $\ell\nu jj$ decay channels with the ATLAS detector at $\sqrt{s} = 7$ TeV. Phys. Rev. D **87**(11), 112006 (2013). doi:10. 1103/PhysRevD.87.112006. arXiv:1305.0125
30. ATLAS Collaboration, Search for production of WW/WZ resonances decaying to a lepton, neutrino and jets in pp collisions at $\sqrt{s} = 8$ TeV with the ATLAS detector. Eur. Phys. J. C **75**(5), 209 (2015). doi:10.1140/epjc/s10052-015-3593-4, doi:10.1140/epjc/s10052-015-3425-6. arXiv:1503.04677 [Erratum: Eur. Phys. J. **C75**, 370 (2015)]
31. CMS Collaboration, Search for massive resonances decaying into pairs of boosted bosons in semi-leptonic final states at $\sqrt{s} = 8$ TeV. JHEP **1408**, 174 (2014). doi:10.1007/ JHEP08(2014)174. arXiv:1405.3447
32. A.L. Read, Presentation of search results: the CL(s) technique. J. Phys. G **28**, 2693–2704 (2002). doi:10.1088/0954-3899/28/10/313
33. T. Junk, Confidence level computation for combining searches with small statistics. Nucl. Instrum. Methods A **434**, 435–443 (1999). doi:10.1016/S0168-9002(99)00498-2. arXiv:hep-ex/9902006
34. J.F. Gunion et al., The Higgs hunter's guide. Front. Phys. **80**, 1–448 (2000)
35. CMS Collaboration, Search for a pseudoscalar boson decaying into a Z boson and the 125 GeV Higgs boson in $\ell^+\ell^-b\bar{b}$ final states. Phys. Lett. B **748**, 221–243 (2015). doi:10.1016/j. physletb.2015.07.010. arXiv:1504.04710
36. ATLAS Collaboration, Search for a CP-odd Higgs boson decaying to Zh in pp collisions at $\sqrt{s} = 8$ TeV with the ATLAS detector. Phys. Lett. B **744**, 163–183 (2014). doi:10.1016/j. physletb.2015.03.054. arXiv:1502.04478

37. ATLAS Collaboration, Search for a new resonance decaying to a W or Z boson and a Higgs boson in the $\ell\ell/\ell\nu/\nu\nu + b\bar{b}$ final states with the ATLAS detector. Eur. Phys. J. C **75**(6), 263 (2015). doi:10.1140/epjc/s10052-015-3474-x. arXiv:1503.08089

38. CMS Collaboration, Search for narrow high-mass resonances in proton-proton collisions at $\sqrt{s}$ = 8 TeV decaying to a Z and a Higgs boson. Phys. Lett. B **748**, 255–277 (2015). doi:10.1016/j.physletb.2015.07.011. arXiv:1502.04994

39. CMS Collaboration, Search for a massive resonance decaying into a Higgs boson and a W or Z boson in hadronic final states in proton-proton collisions at $\sqrt{s}$ = 8 TeV, arXiv:1506.01443

40. ATLAS Collaboration, Search for a charged Higgs Boson produced in the vector-boson fusion mode with decay $H^{\pm} \rightarrow W^{\pm}Z$ using pp collisions at $\sqrt{s}$ = 8 TeV with the ATLAS experiment. Phys. Rev. Lett. **114**(23), 231801 (2015). doi:10.1103/PhysRevLett.114.231801. arXiv:1503.04233

41. CMS Collaboration, Search for resonant pair production of Higgs bosons decaying to two bottom quark-antiquark pairs in proton-proton collisions at 8 TeV. Phys. Lett. B **749**, 560–582 (2015). doi:10.1016/j.physletb.2015.08.047. arXiv:1503.04114

42. ATLAS Collaboration, Search for Higgs boson pair production in the $b\bar{b}b\bar{b}$ final state from pp collisions at $\sqrt{s}$ = 8 TeV with the ATLAS detector. Eur. Phys. J. C **75**, 412 (2015). doi:10.1140/epjc/s10052-015-3628-x. arXiv:1506.00285

43. ATLAS Collaboration, Search for Higgs Boson pair production in the $\gamma\gamma b\bar{b}$ final state using pp collision data at $\sqrt{s}$ = 8 TeV from the ATLAS detector. Phys. Rev. Lett. **114**(8), 081802 (2015). doi:10.1103/PhysRevLett.114.081802. arXiv:1406.5053

Chapter 8
Nonresonant Multi-Boson Production

The observed H(125) boson is broadly compatible with SM expectations. Testing this compatibility with more precision may on the one hand take the form of increasingly precise measurements at the 125 GeV peak. On the other hand, it is possible to verify more indirectly whether the newly discovered boson does indeed play the role it is assigned in the SM. Due to the intimate relationship between the Higgs boson and the W and Z boson in the SM, there are opportunities to observe deviations from the SM especially in the triple and quartic gauge couplings (TGCs and QGCs) of the gauge bosons, which can be observed in the production of multiple bosons.

A common way of searching new physics in multi-boson production in a model-independent way is via an effective field theory. Additional operators are added to the SM Lagrangian, starting with the lowest permitted mass dimensions. The SM itself contains all dimension 4 operators compatible with its symmetries. Higher dimension operators gain coupling factors of $1/\Lambda$ with a dimension of one over mass for each mass dimension above 4. The additional operators can be viewed as an effective expansion of unknown new physics at a mass scale Λ in a low energy approximation around the SM in powers of $1/\Lambda$. The additional operators may induce additional couplings between multiple gauge bosons, accordingly labeled anomalous TGCs or QGCs (aTGCs or aQGCs). Only one dimension 5 operator is possible under these constraints [1], though it is not relevant for hadron collider studies and mostly of interest in neutrino physics. This approach closely follows similar methods used in the context of hadron scattering [2, 3]. In this approach the effective Lagrangian is built on the meson and baryon fields, with the additional terms parameterizing the hadron structure.

In the most general approach, the effective Lagrangian will contain many terms unrelated to vector bosons, which we will not consider here. Even just considering anomalous couplings involving vector bosons leaves a large number of possible operators. For experimental studies, this presents an opportunity as well as problems: different models of physics beyond the SM will lead to deviations from the SM Lagrangian in different couplings, so that the observation of anomalous couplings

M.U. Mozer, *Electroweak Physics at the LHC*, Springer Tracts in Modern Physics 267, DOI 10.1007/978-3-319-30381-9_8 may point to the type of new physics. This advantage makes direct searches in diboson final states attractive compared to indirect studies through precision measurements of the W boson mass, which would be modified by additional diagrams similar to Fig. 5.2. On the other hand it is generally not feasible to analyze a given dataset for the presence of all contributing anomalous couplings simultaneously, taking into account all correlations, so that usually studies are performed under the assumption that only one or two anomalous coupling are present.

The additional terms in the Lagrangian will commonly lead to scattering cross sections rising with the center-of-mass energy s to the point, where the scattering probability becomes larger than one, the so called unitarity limit. This effect is similar to the unitarity violations seen in the Fermi theory of the electroweak interaction: the point-interaction of the Fermi theory is an effective low-energy approach to the full electroweak theory and becomes unphysical in a regime where the finite masses of the W and Z bosons cannot be treated as arbitrarily large any more. Similarly, the unitarity violations computed for anomalous couplings arise in a kinematic regime where the effective field theory is of questionable validity and the unphysical results are presumably cured when the full exotic theory is taken into account.

There is no widely accepted method to suppress the unphysical rise of the cross sections in the absence of the full theory. Ad-hoc form factors [4] or the so-called K- and T-matrix schemes [5, 6] have been used in the past, but also results without any regularization have been published, comparing the experimental cross section limits to bounds derived from unitarity conditions. While these regularization methods provide a way for model-independent studies in the face of divergences, they also inhibit the comparison between results obtained with different regularization methods. Traditionally, the regularization has been viewed as the effect of the undiscovered new physics at a scale not directly accessible. However, at the energies of the LHC and especially when studying strongly divergent quartic couplings, the new physics scale indicated by the regularization schemes are already within experimental reach of the LHC. So the question arises why the new physics responsible for the regularization has not yet been discovered. The issue is confounded, especially for the strongly divergent quartic couplings, by the relatively low sensitivity of the Run I results, where observable effects would imply cross sections near or above the unitarity bound, complicating the interpretation. The issue may be avoided by performing analyses in kinematic regions with well defined upper bounds on the effective center-of-mass energy, so that the interpretation can be confined to new phyiscs scales Λ that are well separated from the experimentally probed region. While this approach is conceptually cleaner than the introduction of regularization methods, it reduces the experimental sensitivity by excluding parts of the experimental data and has not been used in the analyses discussed below.

Non-resonant multi-boson production is also one of the few processes where photon induced processes, as discussed in Sect. 2.2, play a significant role. In fact, photon induced processes are of special interest, as they give direct access to gauge couplings with two photons, as discussed below in the context of exclusive W pair production.

8.1   aTGCs and Dibosons

In the SM, only the WWγ and WWZ vertex are allowed, while the triple neutral couplings are forbidden. This is not necessarily the case for the anomalous couplings. The dimension 6 operators for effective field theories extending the SM have been classified in Refs. [7, 8]. However, not all of these operators are related to TGCs. Among the operators connected to TGCs, a number violates well established conservation laws, such as electric charge conservation, related to electromagnetic gauge invariance. Assuming Lorentz and electromagnetic gauge invariance a simplified set of effective operators emerges [9, 10], leading to eight independent aTGC parameters. Though the parameterization of operators used in the expansion for anomalous couplings is not unique to the above, the notation of Refs. [9, 10] is widely used, easing the comparison of the results. While it should be possible to compare results obtained using different operator bases by using the known relationships between the operators, it is not always possible in practice, as for example the application of unitarization procedures or form factors may spoil the correspondence.

The first direct studies of the TGCs (i.e. processes containing the TGCs at tree level) were performed during the second run period of LEP [11], when the LEP accelerator was running at beam energies high enough to produce W and Z pairs. Measurements of the W pair production cross sections show with high significance that the TGCs are present and that the coupling strengths are compatible with their SM values (see Fig. 8.1).

At the LHC, anomalous triple gauge couplings can similarly be probed in the production of vector boson pairs, i.e. WW, ZZ, WZ, Wγ, and Zγ production (see Fig. 8.2, left). Due to the structure of the anomalous couplings in the Lagrangian, their effects are most visible at high scales, i.e. high invariant masses of the diboson system, high boson p_T, or high p_T of the boson decay products. This makes the

Fig. 8.1 W pair production cross section as function of the center-of-mass energy at LEP, indicating the presence of the WWZ vertex (Adapted from Ref. [11].)

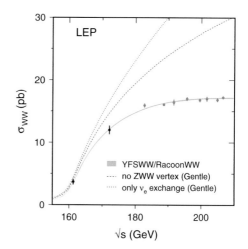

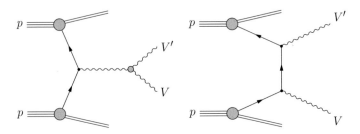

Fig. 8.2 Representative diagrams for diboson production at the LHC, involving (*left*) or missing (*right*) triple gauge interactions. V, V' = W/Z/γ

LHC with its high beam energy especially sensitive to the anomalous couplings. Measurements of the SM couplings, on the other hand, suffer from much larger backgrounds than in the earlier LEP studies, and are not in the main focus of the LHC experiments.

Backgrounds fall into several broad categories. Clearly separable from the signal process are reducible backgrounds from misidentification or misreconstruction. This type of background is especially relevant in studies using hadronic decays or photons. Other backgrounds may have the same initial and final states as the (a)TGC induced processes, so a proper separation is not possible due to interference of the diagrams. Nevertheless, it is usually possible to find a kinematic regime where the (a)TGC contribution to the cross section dominates over other contributions. These irreducible background processes fall into several categories:

- Independent production of two bosons without the involvement of a triple gauge vertex (see Fig. 8.2, right). These processes generally have a much larger total cross sections than the TGC induced processes.
- $t\bar{t}$ production may be viewed as a special case of the above: the independent production of two W bosons. This background may be suppressed by a veto on the presence of b-tagged jets.
- In diboson studies where one of the bosons is a photon, initial and final state radiation off charged particles in the process contribute to the background.

In addition, the SM TGCs appear as background in the search for aTGCs, so that studies of the couplings forbidden in the SM tend to be more sensitive than searches for aTGCs that correspond to allowed triple couplings.

The number of possible boson pairs in combination with the variety of decay modes for the heavy bosons produces a large number of different experimental signatures. Analysis in the all-leptonic decay channels generally have the smallest backgrounds and are thus most suitable in the study of the SM couplings. However, these channels are limited by their low branching ratios in the search for anomalous couplings, where semi-leptonic decays are more promising. The LHC experiments have studied many of these final states, but the set is far from complete. Results are available for ZZ in the all-leptonic final state [12–14], as well as the $2\ell2\nu$ final state [14, 15]. Pairs of heavy bosons have been studied in the semi-leptonic decay

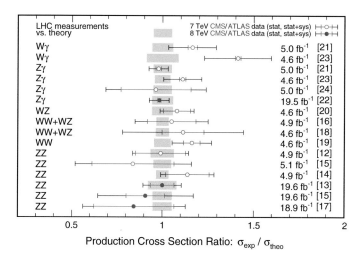

Fig. 8.3 Measured diboson production cross sections compared to the SM expectations. *Points* show experimental measurements (*red* CMS, *blue* ATLAS), while the *yellow bands* represent theory predictions

channel [16–18], as well as the fully leptonic decays [19, 20]. Additionally, measurements of heavy boson production in association with photons have been used to set limits on anomalous couplings with leptonic boson decays [21–23], but also the $Z \rightarrow \nu \bar{\nu}$ decay [23, 24]. No significant deviations from the SM expectations are observed, as shown in Fig. 8.3.

As an example, we will here discuss an analysis in the all-leptonic decays of Z pairs [13]. Similar to the Higgs boson search in the same final state (see Sect. 6), this analysis suffers from a very low branching fraction of $\sim 0.5\%$, when considering electrons and muons. The study also includes τ final states to boost the branching fraction, but the complications of hadronic τ reconstruction and the escaping neutrinos mean that the final contribution of Z decays into electrons and muons dominate the final result. Even though the branching fraction in this final state is low, competitive results are achieved because the final event sample is practically background free. The required presence of four leptons drastically suppresses backgrounds with misidentified leptons and and invariant mass requirement on lepton pairs removes any processes that do not have two real Z bosons in the final state. The absence of tri-neutral SM TGCs leaves only the independent production of two Z bosons as background, which has a rather low cross section. Limits on aTGCs are ultimately extracted from the invariant mass spectrum of the Z pairs, which is expected to show deviations from the SM prediction at high masses in the presence of aTGCs, but no such deviation is observed (see Fig. 8.4).

Fig. 8.4 Spectrum of the di-boson invariant mass in Z pair events with leptonic decays. The observed data matches the SM well (*filled histograms*), but is incompatible with sizable aTGCs (*red line*) (Adapted from Ref. [13].)

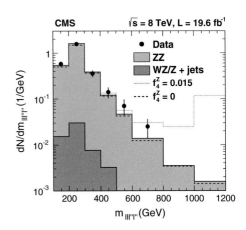

8.2 Quartic Couplings

The approach to the aQGCs is conceptually similar to the triple couplings, with some notable differences. Overall, there is a larger number of possible operators due to the increased number of possible combinations, complicating potential studies. So far, there is also no widely accepted consensus on the operator basis used in computations and measurements concerned with the quartic couplings, partly because relevant results have only been published in the recent past. However, at the LHC an increasing number of results constraining the quartic couplings are being produced and a first extension of the parameterization for aQGCs following the one discussed above for aTGCs has been introduced in [25].

Similar to the case of the triple couplings, the all-neutral quartic couplings are forbidden in the SM, leaving the vertex with four W bosons and the vertices with two W and two neutral bosons. The quartic vertices are of particular interest: in the SM, the rising cross section for the scattering of longitudinally polarized V bosons is regularized by negative interference with diagrams involving the Higgs boson [26–28]. In the absence of the Higgs boson (or other regularizing mechanism), violations of the unitarity bound are expected at a center of mass energy of about 1 TeV. The exact cancellation in the SM is connected to the mechanism of electroweak symmetry breaking: the longitudinal components of the electroweak bosons derive directly from the Higgs field, so that the diagrams shown in Fig. 8.5 in the example of same sign W boson scattering have exact negative interference. While a suitable Higgs boson candidate has been found, anomalies in the longitudinal scattering cross sections may point new physics in the Higgs sector that may not be visible in studies of the H(125) peak [29–31], such as a a composite Higgs boson or new high-mass resonances. Whatever the exact type of new physics involved, its appearance in the vector-boson scattering process would point to an involvement of the process of electroweak symmetry breaking.

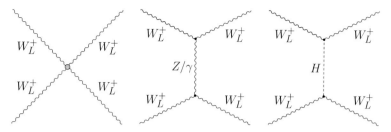

Fig. 8.5 Leading diagrams involved in the scattering of same charge longitudinally polarized W bosons in the SM. The quartic coupling induced amplitude alone would lead to unitarity violations at high energies, but negative interference keeps the total cross section regular

A number of different scenarios have been envisioned that would lead to anomalous gauge couplings. Early studies often involved scenarios that posed alternative electroweak symmetry breaking mechanisms which would explain the boson masses without a Higgs boson, such as the technicolor models [32], where the masses arise as binding energy of a new strong interaction. As the amplitude of the quartic vertex in the longitudinal boson scattering cannot be regularized by diagrams involving the Higgs bosons, visible effects would be expected in processes involving the triple and quartic couplings [33, 34]. Alternatively, the electroweak symmetry breaking may be related to the existence of additional spacial dimensions [35]. In their basic implementations these models do not contain a Higgs boson like particle either, so that such models have fallen out of favor since the discovery of the H(125) boson. However, in more sophisticated implementations Higgs-boson like particles can arise in a scenario called partial compositeness [36, 37]. Doing away with extra dimensions, models can also be constructed where the Higgs boson is not elementary but composite [38, 39]. Even though typical scales of these models are out of direct reach of the LHC, they induce distinct patterns in the anomalous couplings that would allow the LHC experiments to distinguish between broad classes of models based on deviations from the SM in multi-boson measurements alone.

The quartic couplings can be constrained by measurements of triple boson production, similar to the tests of aTGCs in diboson production. Additionally, the QGCs play a role in vector boson scattering (VBS) processes, which are similar to the VBF production process of the Higgs boson (see Chap. 6).

8.2.1 Results of Triple Boson Production

The production cross section for triple boson final states are even lower than diboson final states. Especially the triple production of the massive (W or Z) bosons is suppressed due to a reduction in phase space due to the large amount of energy bound in the rest mass of final state particles. For this reason, only results involving at least one photon are available from Run I of the LHC.

By analyzing events with one leptonic W boson, one photon and one hadronic boson candidate, the CMS experiment has put limits on anomalous WW$\gamma\gamma$ and WWZγ couplings [25]. The choice to study one hadronically decaying boson is driven by the requirement to obtain a reasonably high branching ratio for this rare process, though at least one leptonic decay has to be required as well in order to guarantee a reasonable trigger efficiency and suppression of QCD processes with photons. While inclusion of a hadronically decaying V increases the visible signal, it complicates the interpretation of the results, as the di-jet mass resolution does not suffice to distinguish W and Z bosons, so that two different QGCs contribute to the observation. Even thought the analysis obtains stringent limits on aQGCs, the background from W bosons produced in association with jets and photons in QCD processes is too high to observe the SM QGCs.

The WW$\gamma\gamma$ vertex alone has been studied by the ATLAS experiment by measuring the W$\gamma\gamma$ cross section [40] in events with one lepton, two photons and $E_\mathrm{T}^\mathrm{miss}$. The W$\gamma\gamma$ process is observed above significant background from processes where jets are misidentified as photons, though the cross section is dominated by the independent production, i.e. not involving the quartic vertex. Limits on aTGCs are extracted from the spectrum of photon pair invariant mass, as the aTGCs would lead to an excess at high invariant masses where backgrounds are small.

8.2.2   Vector Boson Scattering Results

The vector boson scattering (VBS) process proceeds via the diagram shown in Fig. 8.6 (top left), leading to a final state of two electroweak bosons in association with two

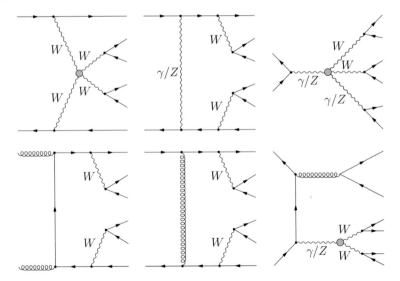

Fig. 8.6 Representative diagrams resulting in WW production in association with two jets. Purely electroweak processes are shown in the *top row*, while the *bottom row* shows mixed QCD electroweak diagrams

jets. Predictions for this channel are complicated by the large number of diagrams leading to the same final state even at LO (see Fig. 8.6 for a representative subset). As the amplitudes interfere with each other, the contributions cannot be easily separated. Nevertheless, it is possible to group the amplitudes into self-consistent sets by counting the powers of the electroweak and strong coupling constants (α_{em} and α_S, respectively). At leading order diagrams proportional to α_{em}^6 (the electroweak component) and $\alpha_{em}^4 \alpha_S^2$ (the QCD component) contribute. Note that the QCD component receives contributions from $t\bar{t}$ production (i.e. Fig. 8.6 bottom left), which is commonly suppressed in studies of this final state by a veto on b-tagged jets. While interference persists even when in this grouping, calculations show that the effect is of the order of only $\sim 10\%$ for the typical kinematic selections used in VBS studies [41]. Only the electroweak component receives contributions from aQGCs, so that the QCD induced processes are usually regarded as background. Although the QCD component has a larger total cross section than the electroweak one, the electroweak processes in general and the VBS process specifically can be enriched by suitable kinematic selections, similar to the selections used to enrich vector-boson-fusion production processes in the Higgs searches (see Sect. 6.5). Nevertheless, the cross section for VBS processes is suppressed in the LHC Run I, as the effective center-of-mass energy of the VV system is substantially lower than the center-of-mass energy of the pp system (7/8 TeV).

Due to the low cross sections of these processes, initial LHC analysis have focused on preparatory studies similar in kinematics to but lacking in contributions from quartic vertices. The goal of these studies is to hone the analysis techniques to be used in the study of VBS processes the first of which will be dicsussed below. The VBS studies will likely gain in importance in Run II, where the increased center-of-mass energy leads to increased cross sections of the VBS processes.

References [42, 43] describe such analyses, measurements of Z boson production in association with two forward jets. The final state receives contribution from TGC induced processes, similar to the one shown in Fig. 8.6 (top left), but replacing the quartic vertex with a triple vertex. A leptonically decaying Z boson, reconstructed as outlined in Sect. 3.1, and two jets are required to be present in the events. Using only this generic selection, the samples are largely dominated by the QCD component. The VBS process is enriched compared to other processes leading to Z + dijet finals states by imposing kinematic requirements on the dijet system: similar to the VBF processes discussed in in Chap. 6, the dijet system in VBS processes is widely separated in pseudorapidity and has a large invariant mass.

A first result on quartic vertices in VBS was obtained in exclusive photoproduction of WW pairs [44], which probes the same WW$\gamma\gamma$ vertex as the triple boson study discussed above. Two leptonically decaying W bosons are identified as described in Sect. 3.1 and the photoproduction processes is enriched by selecting events that do not contain fragmentation products of the initial state protons. This signature is connected to the absence of color- and electric charge flow between the initial state protons, which occurs in most other W pair production processes. Accordingly, the analysis achieves a very pure sample of signal events, resulting in more stringent limits in anomalous WW$\gamma\gamma$ couplings than Ref. [25] even though the results are derived

Fig. 8.7 Invariant mass
spectrum of the two
associated jets in $W^{\pm}W^{\pm}$
production, indicating the
presence of VBS processes
at high masses (Adapted
from Ref. [41].)

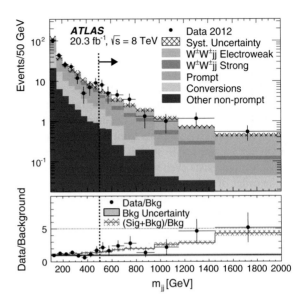

from a data set of smaller integrated luminosity. The veto on proton fragmentation products is very sensitive to the amount of pile-up, as the products of the additional collisions can easily be mistaken for fragmentation products, so that this type of analysis becomes increasingly difficult with increasing LHC luminosity.

A first glimpse of the scattering of heavy bosons may have been seen in a study of W pairs in association with two forward jets [41, 45]. In addition to the two lep-tonically decaying W bosons, two jets with a large difference in η and large invariant mass. While this selection already enriches VBS like processes, the contribution from the electroweak component can be made particularly pure by reducing the analysis to W bosons of the same charge. This requirement reduces the number of contribut-ing diagrams, especially all gluon induced processes, while the VBS contribution is not affected. Figure 8.7 shows the invariant mass distribution for the dijet system in data compared to background estimates, indicating that the observed spectrum is compatible with a substantial VBS contribution to the cross section at high mass. Evidence of the presence of this process is observed with more than 3σ significance, and the corresponding cross section is compatible with the SM.

Even using the full integrated luminosity of the 8 TeV LHC run ($\sim$20 fb^{-1}), the study is still limited by statistical uncertainties. The higher instantaneous luminosity and center-of-mass energy of future LHC runs promise to improve the situation and allow for precision measurements in this channel. The reduced statistical uncertain-ties of Run II may make it possible to study the polarization of the final state bosons, giving preference to the more interesting case of longitudinal boson scattering. The boson polarization is accessible through the W decay asymmetry. However, due to the presence of two neutrinos in the all leptonic VBS final state, it is difficult to reconstruct the decay angles. Alternatively, WZ scattering could be studied, which

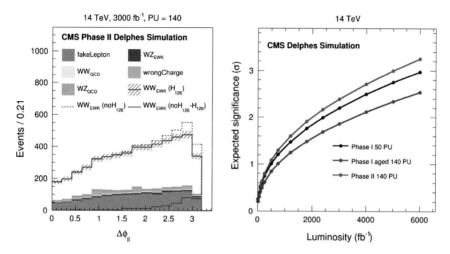

Fig. 8.8 *Left* Opening angle between the two outgoing leptons in a same sign VBS study as expected for the HL-LHC for the SM as well as a Higgsless scenario. *Right* Expected significance for the measurement of the longitudinally polarized component of WW scattering in the fully leptonic same sign channel as function of integrated luminosity for different detector scenarios at the HL-LHC (Adapted from Ref. [46].)

would allow a full final state reconstruction, including the boson decay asymmetries. However, this process has a similarly low cross section as the same-sign WW scattering and the QCD component of the process is not as easily suppressed, so that even the integrated luminosity of the LHC Run II may not suffice. Initial studies of the CMS collaboration show that a major luminosity update of the LHC would make much more detailed measurements possible [46] (see Fig. 8.8). Alternatively, WW scattering could be studied in the semi-leptonic channel, which also gives access to the final state polarization by means of a complete reconstruction of the final state. However, the charge of the hadronically decaying W boson cannot be reliably determined, so that this channel will have significant contributions from the QCD component, as the same-charge process cannot be isolated. Due to the insufficient mass resolution in the reconstruction of the hadronically decaying W, WW and WZ production cannot be well separated. In addition, the production of single W bosons in association with jets is likely to contribute significantly as background. In a SM scenario, the fully leptonic same charge channel is likely to give the optimal performance for a cross section measurement. The semi-leptonic channel, on the other hand, is attractive for the study of anomalous couplings, which have their effect at the highest WW invariant masses, where backgrounds are kinematically suppressed and the larger branching fraction of the semi-leptonic channel is most important. In these analysis, jet substructure algorithms as decribed in Sect. 3.2 may not only significantly reduce the background, but the access to the subjets may also allow the full kinematic reconstruction of the events, including the decay angles related to the boson polarization.

8.3 Outlook

The upgrade of the LHC to higher beam energies and luminosities opens new perspectives for the studies of multi-boson couplings. The enhancement of diboson production at high diboson invariant masses is especially exciting, as this is the kinematic region where contributions from the triple and quartic gauge couplings are most prominent. Measurements of the quartic couplings in particular promise insight into the mechanism of electroweak symmetry breaking and allow for the discovery of new physics in the electroweak sector. However, the cross sections of the most interesting processes are very low, so that the most exciting results will most likely be published only after several years of data-taking.

References

1. S. Weinberg, Baryon and lepton nonconserving processes. Phys. Rev. Lett. **43**, 1566–1570 (1979). doi:10.1103/PhysRevLett.43.1566
2. S. Weinberg, Nonlinear realizations of chiral symmetry. Phys. Rev. **166**, 1568–1577 (1968). doi:10.1103/PhysRev.166.1568
3. J. Gasser, H. Leutwyler, Chiral perturbation theory: expansions in the mass of the strange quark. Nucl. Phys. B **250**, 465 (1985). doi:10.1016/0550-3213(85)90492-4
4. O.J. Eboli et al., Anomalous quartic gauge boson couplings at hadron colliders. Phys. Rev. D **63**, 075008 (2001). doi:10.1103/PhysRevD.63.075008. arXiv:hep-ph/0009262
5. A. Alboteanu, W. Kilian, J. Reuter, Resonances and unitarity in weak boson scattering at the LHC. JHEP **0811**, 010 (2008). doi:10.1088/1126-6708/2008/11/010. arXiv:0806.4145
6. W. Kilian et al., High-energy vector boson scattering after the Higgs discovery. Phys. Rev. D **91**, 096007 (2015). doi:10.1103/PhysRevD.91.096007. arXiv:1408.6207
7. W. Buchmüller, D. Wyler, Effective Lagrangian analysis of new interactions and flavor conservation. Nucl. Phys. B **268**, 621–653 (1986). doi:10.1016/0550-3213(86)90262-2
8. B. Grzadkowski et al., Dimension-six terms in the Standard Model Lagrangian. JHEP **1010**, 085 (2010). doi:10.1007/JHEP10(2010)085. arXiv:1008.4884
9. G. Gounaris, J. Layssac, F. Renard, Signatures of the anomalous $Z\gamma$ and ZZ production at the lepton and hadron colliders. Phys. Rev. D **61**, 073013 (2000). doi:10.1103/PhysRevD.61. 073013. arXiv:hep-ph/9910395
10. G. Gounaris, J. Layssac, F. Renard, Off-shell structure of the anomalous Z and γ selfcouplings. Phys. Rev. D **62**, 073012 (2000). doi:10.1103/PhysRevD.62.073012. arXiv:hep-ph/0005269
11. The ALEPH, DELPHI, L3, OPAL Collaborations, the LEP Electroweak Working Group, Electroweak measurements in electron-positron collisions at W-boson-pair energies at LEP. Phys. Rept. **532**, 119 (2013). doi:10.1016/j.physrep.2013.07.004. arXiv:1302.3415
12. CMS Collaboration, Measurement of the ZZ production cross section and search for anomalous couplings in 2l 2l' final states in pp collisions at $\sqrt{s} = 7$ TeV. JHEP **1301**, 063 (2013). doi:10. 1007/JHEP01(2013)063. arXiv:1211.4890
13. CMS Collaboration, Measurement of the $pp \rightarrow ZZ$ production cross section and constraints on anomalous triple gauge couplings in four-lepton final states at $\sqrt{s} =8$ TeV. Phys. Lett. B **740**, 250–272 (2015). doi:10.1016/j.physletb.2014.11.059. arXiv:1406.0113
14. ATLAS Collaboration, Measurement of ZZ production in pp collisions at $\sqrt{s} = 7$ TeV and limits on anomalous ZZZ and $ZZ\gamma$ couplings with the ATLAS detector. JHEP **1303**, 128 (2013). doi:10.1007/JHEP03(2013)128. arXiv:1211.6096

15. CMS Collaboration, Measurements of the ZZ production cross sections in the $2\ell2\nu$ channel in proton-proton collisions at $\sqrt{s} = 7$ and 8 TeV and combined constraints on triple gauge couplings. Eur. Phys. J. C **75**(10), 511 (2015). doi:10.1140/epjc/s10052-015-3706-0. arXiv:1503.05467

16. CMS Collaboration, Measurement of the sum of WW and WZ production with W+dijet events in pp collisions at $\sqrt{s} = 7$ TeV. Eur. Phys. J. C **73**(2), 2283 (2013). doi:10.1140/epjc/s10052-013-2283-3. arXiv:1210.7544

17. CMS Collaboration, Measurement of WZ and ZZ production in pp collisions at $\sqrt{s} = 8$ TeV in final states with b-tagged jets. Eur. Phys. J. C **74**(8), 2973 (2014). doi:10.1140/epjc/s10052-014-2973-5. arXiv:1403.3047

18. ATLAS Collaboration, Measurement of the $WW + WZ$ cross section and limits on anomalous triple gauge couplings using final states with one lepton, missing transverse momentum, and two jets with the ATLAS detector at $\sqrt{s} = 7$ TeV. JHEP **1501**, 049 (2015). doi:10.1007/JHEP01(2015)049. arXiv:1410.7238

19. ATLAS Collaboration, Measurement of W^+W^- production in pp collisions at $\sqrt{s}$=7 TeV with the ATLAS detector and limits on anomalous WWZ and WWγ couplings. Phys. Rev. D **87**(11), 112001 (2013). doi:10.1103/PhysRevD.87.112001, doi:10.1103/PhysRevD.88.079906. arXiv:1210.2979

20. ATLAS Collaboration, Measurement of WZ production in proton-proton collisions at $\sqrt{s} = 7$ TeV with the ATLAS detector. Eur. Phys. J. C **72**, 2173 (2012). doi:10.1140/epjc/s10052-012-2173-0. arXiv:1208.1390

21. CMS Collaboration, Measurement of the $W\gamma$ and $Z\gamma$ inclusive cross sections in pp collisions at $\sqrt{s} = 7$ TeV and limits on anomalous triple gauge boson couplings. Phys. Rev. D **89**(9), 092005 (2014). doi:10.1103/PhysRevD.89.092005. arXiv:1308.6832

22. CMS Collaboration, Measurement of the Zγ production cross section in pp collisions at 8 TeV and search for anomalous triple gauge boson couplings. JHEP **1504**, 164 (2015). doi:10.1007/JHEP04(2015)164. arXiv:1502.05664

23. ATLAS Collaboration, Measurements of Wγ and Zγ production in pp collisions at $\sqrt{s} = 7$ TeV with the ATLAS detector at the LHC. Phys. Rev. D **87**(11), 112003 (2013). doi:10.1103/PhysRevD.87.112003. arXiv:1302.1283

24. CMS Collaboration, Measurement of the production cross section for $Z\gamma \rightarrow \nu\bar{\nu}\gamma$ in pp collisions at $\sqrt{s} = 7$ TeV and limits on $ZZ\gamma$ and $Z\gamma\gamma$ triple gauge boson couplings. JHEP **1310**, 164 (2013). doi:10.1007/JHEP10(2013)164. arXiv:1309.1117

25. CMS Collaboration, A search for WWγ and WZγ production and constraints on anomalous quartic gauge couplings in pp collisions at $\sqrt{s} = 8$ TeV. Phys. Rev. D **90**, 032008 (2014). doi:10.1103/PhysRevD.90.032008. arXiv:1404.4619

26. M. Veltman, Second threshold in weak interactions. Acta Phys. Polon. B **8**, 475 (1977)

27. B.W. Lee, C. Quigg, H. Thacker, The strength of weak interactions at very high-energies and the Higgs boson mass. Phys. Rev. Lett. **38**, 883–885 (1977). doi:10.1103/PhysRevLett.38.883

28. B.W. Lee, C. Quigg, H. Thacker, Weak interactions at very high-energies: the role of the Higgs boson mass. Phys. Rev. D **16**, 1519 (1977). doi:10.1103/PhysRevD.16.1519

29. D. Espriu, B. Yencho, Longitudinal WW scattering in light of the Higgs boson discovery. Phys. Rev. D **87**(5), 055017 (2013). doi:10.1103/PhysRevD.87.055017. arXiv:1212.4158

30. J. Chang et al., WW scattering in the era of post-Higgs-boson discovery. Phys. Rev. D **87**(9), 093005 (2013). doi:10.1103/PhysRevD.87.093005. arXiv:1303.6335

31. A. Ballestrero et al., How well can the LHC distinguish between the SM light Higgs scenario, a composite Higgs and the Higgsless case using VV scattering channels? JHEP **0911**, 126 (2009). doi:10.1088/1126-6708/2009/11/126. arXiv:0909.3838

32. T. Appelquist, G.-H. Wu, The electroweak chiral Lagrangian and new precision measurements. Phys. Rev. D **48**, 3235–3241 (1993). doi:10.1103/PhysRevD.48.3235. arXiv:hep-ph/9304240

33. J. Pelaez, Resonance spectrum of the strongly interacting symmetry breaking sector. Phys. Rev. D **55**, 4193–4202 (1997). doi:10.1103/PhysRevD.55.4193. arXiv:hep-ph/9609427

34. A. Dobado et al., CERN LHC sensitivity to the resonance spectrum of a minimal strongly interacting electroweak symmetry breaking sector. Phys. Rev. D **62**, 055011 (2000). doi:10.1103/PhysRevD.62.055011. arXiv:hep-ph/9912224

35. H. Davoudiasl et al., Higgsless electroweak symmetry breaking in warped backgrounds: constraints and signatures. Phys. Rev. D **70**, 015006 (2004). doi:10.1103/PhysRevD.70.015006. arXiv:hep-ph/0312193

36. R. Contino, Y. Nomura, A. Pomarol, Higgs as a holographic pseudo-Goldstone boson. Nucl. Phys. B **671**, 148–174 (2003). doi:10.1016/j.nuclphysb.2003.08.027. arXiv:hep-ph/0306259

37. K. Agashe, R. Contino, A. Pomarol, The minimal composite Higgs model. Nucl. Phys. B **719**, 165–187 (2005). doi:10.1016/j.nuclphysb.2005.04.035. arXiv:hep-ph/0412089

38. D.B. Kaplan, H. Georgi, SU(2) x U(1) breaking by vacuum misalignment. Phys. Lett. B **136**, 183 (1984). doi:10.1016/0370-2693(84)91177-8

39. G. Giudice et al., The strongly-interacting light Higgs. JHEP **0706**, 045 (2007). doi:10.1088/1126-6708/2007/06/045. arXiv:hep-ph/0703164

40. ATLAS Collaboration, Evidence of $W\gamma\gamma$ production in pp collisions at $\sqrt{s} = 8$ TeV and limits on anomalous quartic gauge couplings with the ATLAS detector. Phys. Rev. Lett. **115**(3), 031802 (2015). doi:10.1103/PhysRevLett.115.031802. arXiv:1503.03243

41. ATLAS Collaboration, Evidence for electroweak production of $W^{\pm}W^{\pm}jj$ in pp collisions at $\sqrt{s} = 8$ TeV with the ATLAS detector. Phys. Rev. Lett. **113**(14), 141803 (2014). doi:10.1103/PhysRevLett.113.141803. arXiv:1405.6241

42. CMS Collaboration, Measurement of electroweak production of two jets in association with a Z boson in proton-proton collisions at $\sqrt{s} = 8$ TeV. Eur. Phys. J. C **75**(2), 66 (2015). doi:10.1140/epjc/s10052-014-3232-5. arXiv:1410.3153

43. ATLAS Collaboration, Measurement of the electroweak production of dijets in association with a Z-boson and distributions sensitive to vector boson fusion in proton-proton collisions at $\sqrt{s} = 8$ TeV using the ATLAS detector. JHEP **1404**, 031 (2014). doi:10.1007/JHEP04(2014)031. arXiv:1401.7610

44. CMS Collaboration, Study of exclusive two-photon production of $W^{+}W^{-}$ in pp collisions at $\sqrt{s} = 7$ TeV and constraints on anomalous quartic gauge couplings. JHEP **1307**, 116 (2013). doi:10.1007/JHEP07(2013)116. arXiv:1305.5596

45. CMS Collaboration, Study of vector boson scattering and search for new physics in events with two same-sign leptons and two jets. Phys. Rev. Lett. **114**, 051801 (2015). doi:10.1103/PhysRevLett.114.051801. arXiv:1410.6315

46. J. Butler et al., Technical Proposal for the Phase-II Upgrade of the CMS Detector, Technical Report CERN-LHCC-2015-010. LHCC-P-008, CERN, Geneva. Geneva, (Jun, 2015). Upgrade Project Leader Deputies: Lucia Silvestris (INFN-Bari), Jeremy Mans (University of Minnesota) Additional contacts: Lucia. Silvestris@cern.ch, Jeremy. Mans@cern.ch

Chapter 9
Conclusion

Now, that data-analysis of the first data-taking period of the LHC is coming to a close, we may look back to review the progress this marvelous machine has brought to the particle physics community. After an initially slow start in 2010, the LHC quickly reached very high instantaneous luminosities, accumulating a substantial amount of data in 2011 and 2012. The resulting analysis pushed our knowledge of particle physics to a new frontier, culminating in the discovery of the Higgs boson and thus completing the observation of the particles of the SM.

While the LHC had been explicitly designed for the hunt for the Higgs boson, it has produced an unprecedented number of electroweak bosons and the ATLAS and CMS experiments are perfectly suited to study them. Thus, the LHC should not be solely considered to be a machine built for the discovery of new physics but also as an electroweak boson factory, giving access to the many facets of electroweak physics.

While precision measurements of electroweak parameters are a challenge at the LHC due to the large amount of pile-up and unknown initial state of pp collisions, measurements with electroweak bosons have proven particularly fruitful at the boundary to other topics of particles physics. Especially, the leptonic decays of electroweak bosons serve as distinct probes into otherwise difficult to access properties of QCD. These measurements are continuously improving our understanding of the proton structure, where the flavor sensitivity of the electroweak bosons proves to be particularly useful. Using this improved understanding of QCD and the vector boson production processes, the precision of electroweak parameter measurements may surpass previous results from the Tevatron in the near future. This detailed knowledge of SM processes is also necessary to derive predictions for the backgrounds in searches for new particles or the measurements of rare processes.

The most celebrated rare process studied at the LHC is without doubt the production of the Higgs boson. The intimate link of the Higgs boson to the electroweak bosons is the reason why the search for the Higgs boson was performed to great depth in electroweak boson final states. The discovery in 2012 will be remembered as a milestone for particle physics for the foreseeable future. Already with the limited data collected during the LHC Run I it has been possible to study the properties

© Springer International Publishing Switzerland 2016
M.U. Mozer, *Electroweak Physics at the LHC*, Springer Tracts
in Modern Physics 267, DOI 10.1007/978-3-319-30381-9_9

of the new boson in surprising detail. The results show that the H(125) boson is largely compatible with the SM Higgs boson in quantum numbers and couplings. The increased luminosity expected for the second running period of the LHC is likely to improve the accuracy of these property measurements substantially, opening the new field of precision Higgs physics.

However, the SM is not without problems, failing to explain the neutrino masses and the apparent presence of so called "dark matter" and "dark energy" in astronomical and cosmological surveys. Additionally, the SM requires suspect numerical coincidences to explain the observations, begging the question whether these coincidences my instead be the result of new and undiscovered physical principles. A large number of extensions of the SM have been devised which may resolve this issues, such as supersymmetric theories, extended Higgs sectors or theories involving spatial extra-dimensions. Many of these theories predict additional particles which may be observed at the LHC. For example, studies of electroweak bosons may also reveal the solution to the Hierarchy problem, as beyond SM theories that address the problem by introducing additional spatial dimensions may produce resonances decaying to electroweak bosons. The sensitivity of such searches at the LHC for new particles decaying the vector bosons have greatly profited from the introduction of jet substructure techniques for the reconstruction of energetic boson decays. These techniques allow us to leverage the large hadronic branching fraction of the electroweak bosons to push these studies to new heights of sensitivity, far surpassing naive extrapolations from Tevatron results.

Even after the discovery of the Higgs boson, studies of the mechanism of electroweak symmetry breaking remain relevant, as the deviations from SM expectations may point to new physics. The symmetry breaking may either be studied directly through measurements of the Higgs boson properties, or indirectly. Of special interest is the study of the mechanism of electroweak symmetry breaking by the possible observation of longitudinal W boson scattering, which is a unique prediction of the SMs mechanism of mass generation. This process is studied in the context of non-resonant multi-boson production either in tri-boson production or in di-boson production in association with jets. However, the cross section for these processes is extremely small, so that the LHC Run I has only allowed for a very first glimpse into this interesting sector. Analyses have been more productive in the study of simple non-resonant di-boson production, which is not connected to electroweak symmetry breaking, but allows for the model independent search of new physics through anomalous couplings in terms of effective field theories.

The prospects for future running at the LHC are equally good. The higher luminosity and center of mass energy allow us to probe much higher scales for possible effects of new physics, but also open the gates for detailed study of the multi-boson processes. At the high energies involved boosted decays will become increasingly important, calling for improvements of our reconstruction techniques, considering that current jet-substructure techniques are still in their infancy.

Away from the energy frontier, the improved understanding of the LHC detectors and an increasing number of subsidiary measurements may enable a more precise determination of the W mass. Such a measurement will put the SM to a stringent consistency test when combined with other electroweak precision measurements.

After the current Run II of the LHC, another phase of LHC running may follow, the so called High Luminosity LHC (HL-LHC), intended to reach integrated luminosities of $3000\,\mathrm{fb}^{-1}$ at a beam energy of 7 TeV. The LHC detectors will require significant upgrades, replacing radiation damaged parts and preparing for the much increased pile-up necessary to reach these enormous luminosities. This phase of LHC running would provide access to processes with extremely low cross sections and allow for a very detailed study of VBS processes. Additionally, the very high luminosities promise high precision measurements of the Higgs boson properties, possibly even the Higgs self-couplings. The search for exotic resonances, on the other hand, is mostly driven by the center-of-mass energy and will most likely profit less from this vast increase in luminosity.

Beyond the LHC, future collider projects may push the boundaries of our understanding of electroweak physics even further. A possible future lepton collider may use high statistics precision measurements of the Z and W boson to search indirectly for new physics (see for example [1, 2]). Studies of Higgs bosons in associated production would also allow more detailed studies of the coupling of the Higgs boson to electroweak bosons, further validating the SM or find possible deviations. Alternatively, an improved hadron collider [3] may increase the reach in the search for new physics up to a scale of dozens of TeV. No matter which direction the field as a whole will take, electroweak physics can be expected to remain a central theme and driving force in the future as it has been in the past at $Sp\overline{p}S$, SLC, LEP and the Tevatron.

References

1. J. Erler et al., Physics impact of GigaZ. Phys. Lett. B **486**, 125–133 (2000). doi:10.1016/S0370-2693(00)00749-8. arXiv:hep-ph/0005024
2. S. Heinemeyer, G. Weiglein, Top, GigaZ, MegaW, arXiv:1007.5232
3. VLHC Design Study Group Collaboration, Design study for a staged very large hadron collider, Technical Report SLAC-R-591, SLAC-R-0591, SLAC-591, SLAC-0591, FERMILAB-TM-2149 (2001)

Printed in the United States
By Bookmasters